Sensor Technology for Sports Monitoring

Sensor Technology for Sports Monitoring

Editor

Vesa Linnamo

MDPI • Basel • Beijing • Wuhan • Barcelona • Belgrade • Manchester • Tokyo • Cluj • Tianjin

Editor
Vesa Linnamo
University of Jyväskylä
Jyväskylä
Finland

Editorial Office
MDPI
St. Alban-Anlage 66
4052 Basel, Switzerland

This is a reprint of articles from the Special Issue published online in the open access journal *Sensors* (ISSN 1424-8220) (available at: https://www.mdpi.com/journal/sensors/special_issues/STSM).

For citation purposes, cite each article independently as indicated on the article page online and as indicated below:

LastName, A.A.; LastName, B.B.; LastName, C.C. Article Title. *Journal Name* **Year**, *Volume Number*, Page Range.

ISBN 978-3-0365-7088-4 (Hbk)
ISBN 978-3-0365-7089-1 (PDF)

Contents

About the Editor

Vesa Linnamo

Professor Vesa Linnamo received his Ph.D. in Biomechanics from the University of Jyväskylä, Finland in 2002. His research interests involve motor control and neuromuscular adaptation along with sports biomechanics, especially in Nordic winter sports. He is currently working in the Faculty of Sport and Health Sciences, University of Jyväskylä as a professor in sports technology and the head of the Vuokatti Sports Technology Unit in Vuokatti, Finland.

sensors MDPI

Editorial

Sensor Technology for Sports Monitoring

Vesa Linnamo

Faculty of Sport and Health Sciences, University of Jyväskylä, 40014 Jyväskylä, Finland; vesa.linnamo@jyu.fi; Tel.: +358-40504-4800

Over the past decades, huge steps have been made in the development of sensor technology related to sports monitoring. Sensors are lighter, data transmission is mostly wireless, and software applications are more user-friendly. Wearable technologies have expanded from heart rate monitoring to, e.g., the use of inertial measurement units (IMUs) to detect motion and technique changes in different sports. IMUs combined with satellite systems also make speed detection and the analysis of an athlete's racing performance possible. New technologies have emerged for both laboratory and field conditions. In laboratory, it is easier to create more stable conditions, but if laboratory tests do not correspond to the actual sport performance on the field, one should be careful with the conclusions.

The challenge with any new sensor and/or algorithm is that it needs to be both valid and reliable. In other words, it needs to utilize applicable measures that are accurate and repeatable while additional measurements conditions remain unchanged. A common way to test new systems is to compare them against a "golden standard", which is considered the most accurate available system for that purpose. If the measurement error is higher than the change expected to be seen in the performance variable, naturally, the system is not very useful. In a coaching situation, feedback to the athlete should, however, be given without too much delay and the coach needs to decide whether they are willing to compromise accuracy to save time.

This Special Issue "Sensor Technology for Sports Monitoring" addresses the topics raised above. It consists of ten papers related to all kinds of sensors that are currently being used for monitoring different sports. The topics vary from dancing to winter sports such as figure skating, alpine and cross-country skiing with para-athletic and able-bodied skiers, to summer sports such as football and kayaking.

The aim of the first article [1] in this Special Issue was to test the performance of an enhanced version of a prototype in a dynamic skiing-like bending simulation as well as in a proof-of-concept field measurement. It was concluded that combined with the high laboratory-based reliability and validity of the PyzoFlex® prototype, it can be a potential candidate for smart ski equipment.

In [2], a system for detecting and visualizing the very important dance motions known as stops was introduced. Using a neural network to learn motion features, the system was able to detect stops and visualize them using a human-like 3D model with highly accurate stop detection results. Its effectiveness as an information and communication technology to support remote group dance practice was demonstrated.

Paper [3] is a pilot study introducing a motion analysis system to monitor alpine skiing activities during training sessions. Through five inertial measurement units (IMUs) placed on five points of the athletes, the angle of each joint was computed to evaluate the ski run. The aim of this work was to find a tool to support ski coaches during training sessions.

The aims of article [4] were to analyze the differences in physical demands of non-starter players regarding playing time during competition and to evaluate the physical demands of the compensatory training (MD + 1C) for substitute players in elite football. The study highlighted the importance of individualizing the workload of training sessions and suggested that the complementary session should be individualized according to the minutes played by the substitutes who are potentially under-loaded compared to starters.

Citation: Linnamo, V. Sensor Technology for Sports Monitoring. *Sensors* **2023**, *23*, 572. https://doi.org/10.3390/s23020572

Received: 7 December 2022
Revised: 7 December 2022
Accepted: 9 December 2022
Published: 4 January 2023

In paper [5], the authors evaluated two approaches for estimating the total pro-pulsive force on a skier's center of mass with double-poling and V2-skating skiing techniques using force measurement bindings, pole force sensors, and an eight-camera Vicon system for data collection. Both approaches properly estimated the trend of the force–time curve of the propulsive force, but due to its better accuracy, it was suggest-ed that an approach that calculates the forward component of the ground reaction force may be a more appropriate method than using the forward component of the translational force.

Article [6] presents the design and experimental evaluation of a non-invasive wearable sensor system that can be used to acquire crucial information about athletes' performance during inline figure skating training. The system was able to detect when the jumps were performed and provided a live view of the data through a graphical user interface. The results confirmed its effectiveness in correctly detecting jumps, especially considering its compromise between precision and the overall cost of the equipment.

Papers [7,8] both studied kayaking. The aim of [7] was to investigate the magnitude and shape of the forces applied on the foot rest, foot strap, and paddle, and it was found that when comparing the best and worst kayakers' performance, the best showed greater magnitudes of force and greater synchronization of the peak forces. It was recommended that analyses of the force–time curves, including not only the forces applied by the kayaker on the paddle but also the ones applied on the foot rest and strap, should be considered relevant in terms of technique analyses. In article [8], the aim was to compare key variables of paddle strokes measured by a commercial Trainesense SmartPaddle® against the strain-gauge shaft and investigate how these variables are associated with the velocity of the boat among national-level canoe polo players. It was concluded that the SmartPaddle can provide promising information on key stroke variables when compared to the strain-gauge paddle shaft. In the future, this could be an interesting tool for biomechanical research and daily kayaking coaching in real open water conditions.

The topic of article [9] was related to para-athletic cross-country (XC) skiing. The paper's aims were to evaluate the feasibility of a framework based on micro-sensor technology for, firstly, in-field analyses of performance and, secondly, sub-technique selection in the field by using it to compare different parameters between elite standing para-athletic and able-bodied XC skiers during a classical skiing race. The data from a global navigation satellite system and inertial measurement unit were integrated to compare time loss and selected sub-techniques as a function of speed. It was found that there can be different speed thresholds in the classical sub-techniques for para-athletic skiers versus able-bodied skiers. It was suggested that the framework could provide a point of departure for large-scale international investigations of performance and related factors in para-athletic XC skiing.

The aim of the final paper [10] of this Special Issue was to develop a novel ski demonstrator and to conceptualize and validate an empirical curvature model. PyzoFlex® technology-based sensor foils were attached to the upper surface of an al-pine ski, and a self-developed instrument was used as a data acquisition device. Although the new measuring system offered good accuracy and high precision, it was suggested that a transfer into the field is only allowed to a limited extent since the scope of the curvature model has not yet been definitely determined. The authors concluded that high laboratory-related reliability and validity of their novel ski prototype featuring PyzoFlex® technology make it a potential candidate for on-snow application, such as smart skiing equipment.

In order to be able to analyze and give proper advice on sport techniques, it is important to understand the biomechanical and physiological demands of different sports. Thus, various sport monitoring systems are needed to help athletes achieve optimal performance. As this Special Issue shows, several developments in this field have been made all over the world in different research institutes. The work, however, continues and should continue to make sensors in this field even more accurate and user-friendly in the future.

Acknowledgments: The Guest Editor is very grateful to all the authors for publishing their valuable research, which led to the success of this Special Issue.

Conflicts of Interest: The author declares no conflict of interest.

References

1. Thorwartl, C.; Kröll, J.; Tschepp, A.; Holzer, H.; Teufl, W.; Stöggl, T. Validation of a Sensor-Based Dynamic Ski Deflection Measurement in the Lab and Proof-of-Concept Field Investigation. *Sensors* **2022**, *22*, 5768. [CrossRef] [PubMed]
2. Jin, Y.; Suzuki, G.; Shioya, H. Detecting and Visualizing Stops in Dance Training by Neural Network Based on Velocity and Acceleration. *Sensors* **2022**, *22*, 5402. [CrossRef] [PubMed]
3. Russo, C.; Puppo, E.; Roati, S.; Somà, A. Proposal of an Alpine Skiing Kinematic Analysis with the Aid of Miniaturized Monitoring Sensors, a Pilot Study. *Sensors* **2022**, *22*, 4286. [CrossRef] [PubMed]
4. Calderón-Pellegrino, G.; Gallardo, L.; Garcia-Unanue, J.; Luis Felipe, J.L.; Hernandez-Martin, A.; Paredes-Hernández, V.; Sánchez-Sánchez, J. Physical Demands during the Game and Compensatory Training Session (MD + 1) in Elite Football Players Using Global Positioning System Device. *Sensors* **2022**, *22*, 3872. [CrossRef] [PubMed]
5. Zhao, S.; Ohtonen, O.; Ruotsalainen, K.; Kettunen, L.; Lindinger, S.; Göpfert, C.; Linnamo, V. Propulsion Calculated by Force and Displacement of Center of Mass in Treadmill Cross-Country Skiing. *Sensors* **2022**, *22*, 2777. [CrossRef] [PubMed]
6. Panfili, A.; Spanò, A.; Cortesi, A. A Wearable System for Jump Detection in Inline Figure Skating. *Sensors* **2022**, *22*, 1650. [CrossRef] [PubMed]
7. Bonito, P.; Sousa, M.; Ferreira, F.M.; Justo, J.F.; Gomes, B.B. Magnitude and Shape of the Forces Applied on the Foot Rest and Paddle by Elite Kayakers. *Sensors* **2022**, *22*, 1612. [CrossRef] [PubMed]
8. Löppönen, A.; Vänttinen, T.; Haverinen, M.; Linnamo, V. The Effect of Paddle Stroke Variables Measured by Trainesense SmartPaddle® on the Velocity of the Kayak. *Sensors* **2022**, *22*, 938. [CrossRef] [PubMed]
9. Carlsen, C.H.; Baumgart, J.K.; Kocbach, J.; Haugnes, P.; Paulussen, E.M.B.; Sandbakk, Ø. Framework for In-Field Analyses of Performance and Sub-Technique Selection in Standing Para Cross-Country Skiers. *Sensors* **2021**, *21*, 4876. [CrossRef] [PubMed]
10. Thorwartl, C.; Kröll, J.; Tschepp, A.; Schäffner, P.; Holzer, H.; Stöggl, T. A Novel Sensor Foil to Measure Ski Deflections: Development and Validation of a Curvature Model. *Sensors* **2021**, *21*, 4848. [CrossRef] [PubMed]

sensors

MDPI

Article

Validation of a Sensor-Based Dynamic Ski Deflection Measurement in the Lab and Proof-of-Concept Field Investigation

Christoph Thorwartl [1,*], Josef Kröll [1], Andreas Tschepp [2], Helmut Holzer [3], Wolfgang Teufl [1] and Thomas Stöggl [1,4]

1 Department of Sport and Exercise Science, University of Salzburg, Schlossallee 49, 5400 Hallein/Rif, Austria; josef.kroell@sbg.ac.at (J.K.); wolfgang.teufl@sbg.ac.at (W.T.); thomas.stoeggl@sbg.ac.at (T.S.)
2 Joanneum Research Forschungsgesellschaft mbH, Franz-Pichler-Straße 30, 8160 Weiz, Austria; andreas.tschepp@joanneum.at
3 Atomic Austria GmbH, Atomic Strasse 1, 5541 Altenmarkt, Austria; helmut.holzer@atomic.com
4 Red Bull Athlete Performance Center, Brunnbachweg 71, 5303 Thalgau, Austria
* Correspondence: christoph.thorwartl@sbg.ac.at; Tel.: +43-664-3455681

Abstract: Introduction: Ski deflection is a performance-relevant factor in alpine skiing and the segmental and temporal curvature characteristics (m^{-1}) along the ski have lately received particular attention. Recently, we introduced a PyzoFlex® ski deflection measurement prototype that demonstrated high reliability and validity in a quasi-static setting. The aim of the present work is to test the performance of an enhanced version of the prototype in a dynamic setting both in a skiing-like bending simulation as well as in a field proof-of-concept measurement. Material and methods: A total of twelve sensor foils were implemented on the upper surface of the ski. The ski sensors were calibrated with an empirical curvature model and then deformed on a programmable bending robot with the following program: 20 times at three different deformation velocities (v_{slow}, v_{medium}, v_{fast}) with (1) central bending, (2) front bending, (3) back bending, (4) edging left, and (5) edging right. For reliability assessment, pairs of bending cycles (cycle 1 vs. cycle 10 and cycle 10 vs. cycle 20) at v_{slow}, v_{medium}, and v_{fast} and between pairs of velocity (v_{slow} vs. v_{medium} and v_{slow} vs. v_{fast}) were evaluated by calculating the change in the mean (CIM), coefficient of variation (CV) and intraclass correlation coefficient (ICC 3.1) with a 95% confidence interval. For validity assessment, the calculated segment-wise mean signals were compared with the values that were determined by 36 infrared markers that were attached to the ski using an optoelectrical measuring system (Qualisys). Results: High reliability was found for pairs of bending cycles (CIM −0.69–0.24%, max CV 0.28%, ICC 3.1 > 0.999) and pairs of velocities (max CIM = 3.03%, max CV = 3.05%, ICC 3.1 = 0.997). The criterion validity based on the Pearson correlation coefficient was r = 0.98. The accuracy (systematic bias) and precision (standard deviation), were $-0.003\ m^{-1}$ and $0.047\ m^{-1}$, respectively. Conclusions: The proof-of-concept field measurement has shown that the prototype is stable, robust, and waterproof and provides characteristic curvature progressions with plausible values. Combined with the high laboratory-based reliability and validity of the PyzoFlex® prototype, this is a potential candidate for smart ski equipment.

Keywords: bending sensors; flexion; PyzoFlex; ski bending; ski deflection

Citation: Thorwartl, C.; Kröll, J.; Tschepp, A.; Holzer, H.; Teufl, W.; Stöggl, T. Validation of a Sensor-Based Dynamic Ski Deflection Measurement in the Lab and Proof-of-Concept Field Investigation. *Sensors* **2022**, *22*, 5768. https://doi.org/10.3390/s22155768

Academic Editor: Michael E. Hahn

Received: 5 July 2022
Accepted: 29 July 2022
Published: 2 August 2022

1. Introduction

The way a ski moves on the snow surface provides information about the performance and quality of a turn. A ski that slides sideways across the snow surface during forward motion is called skidding. In contrast, a carving turn is defined by minimal or no lateral displacement of the ski relative to the track, and, therefore, a point along the ski edge follows the path of the proceeding one [1–4]. To characterize a carving turn, the term "turn radius" has been established, which is defined by side-cut and ski deflection [5,6]. While carving, the turn radius can be changed by the manipulation of the ski deflection based on alterations in the applied skiing technique (e.g., applied forces, edge angle, body position). However, it is an oversimplification to assume a homogeneous deflection behavior (=constant radius) along the ski; rather, temporal and segmental differences in the curvature (w'') occur depending on the skiing technique and the quality of a turn [3,7]. Accordingly, it is not possible to deduce the deflection behavior based on the turn radius; instead, the w'' characteristic along the ski is decisive.

The most common method to measure ski deflection under standardized laboratory conditions is the three-point bending test [8–14]. For analyzing the ski deflection in the field, only strain gauges [6,15–23] have been used. The strain gauge-based approaches have not been tested for reliability or validity from a scientific point of view, and consequently the studies are limited for proof-of-concept investigations. In contrast, a novel PyzoFlex® prototype has already been tested for reliability and validity in a quasi-static setting with a high-precision laser measurement system [7]. A curvature model was developed and the calibrated data provided both good accuracy (1.33×10^{-3} m^{-1}) and high precision (4.14×10^{-3} m^{-1}) [7]. However, no field measurements have been performed with this novel prototype so far.

The calibration of bending sensors is commonly performed in a quasi-static laboratory bending simulation. The resulting bending line of the deformed ski can be quantified by the vertical displacement [8–12], the angular deformation [13], or the curvature (w'') [7,14]. Those values can accordingly be used for sensor calibration. With this procedure, it is assumed that sensors that are calibrated in the quasi-static setting will measure correctly also during dynamic deflections in the field. There are three studies that have gone a step further and not only considered the quasi-static approach, but also verified the sensor signal in dynamics. In the study of Yoneyama et al. (2008), a free vibration measurement was performed, demonstrating good agreement between the signal from the strain gauges and the external inductive displacement transducer [18]. However, the method was limited to a qualitative description and the free vibration is a dynamic process, which is not very similar to the real ski-specific deflections. In the work of Thorwartl et al. (2022), the PyzoFlex® prototype showed reliable and plausible curvature characteristics in a dynamic oscillating bending situation, although the measured curvature was only estimated and not compared with a gold standard measuring device [24]. To the best of our knowledge, the study of Adelsberger et al. (2014) is the only one which performed a validation in the field. The center of mass (CoM) turn radii that was obtained from a differential Global Navigation Satellite System was compared with data from strain gauges. The root mean square error between the radii that were calculated from deflection measurements and CoM radii was on average 1.26 m [6]. Moreover, it was pointed out that the radii were smaller relative to the track radii at the tip of the ski. This indicates an inhomogeneous deflection behavior of the ski, but only the total turn radius and not segment-specific radius differences were validated. Individual sensors were consequently not analyzed to validate the ski deflection.

None of the prototypes that were described above were subjected to a reliability or validity assessment in a dynamic ski-specific deformation setting. Validating the deflection in the field is complex, since we do not know of any gold standard measuring instrument for deflection detection on snow. Therefore, it is appropriate to perform the dynamic validation in the laboratory by deforming the ski using a programmable bending robot applying deformations that are similar to skiing (e.g., cyclic deformation, edging left and right). In addition to the influence of cyclic loads and combined deformations (torsion and bending), the ski would also have to be deformed with different excitation frequencies in order to determine and, if necessary, extend the scope of the deflection model.

By using such a laboratory setup, a gold standard instrument must be used that can detect ski deflection in three-dimensional space. The best accuracies of local position measurement technologies are provided by optoelectronic measurement systems (OMS) and, therefore, they are often considered as the "gold standard" in 3D motion capture [25,26]. The position accuracy of OMS is between 0.01 and 0.77 mm, depending on the capture system and experimental setup (e.g., the number of markers and cameras, marker size, size of the volume to be captured and distances between the cameras) [26–32]. However, the accuracy of the w'' detection can only be estimated to a limited extent from the local position accuracy. Based on measurements with a calibration profile it was shown that no valid w'' can be detected at a marker distance of <50 mm and only conditionally at 110 mm [33]. At a marker distance of 170 mm, valid curvatures can be detected since the maximum average difference was 1.11×10^{-3} m^{-1} in the static situation and -3.83×10^{-3} m^{-1}, in the dynamic situation [33]. In only one study, the radius of curvature was estimated using an OMS for bending stiffness determination of alpine skis, but the w'' was not calculated exactly [12].

Since the recently published study of the novel PyzoFlex® ski prototype was restricted to quasi-static setting, the purpose of the current study was to verify and extend the curvature model with respect to dynamic deflection. Therefore, the aims of this study were to (a) present an enhanced PyzoFlex® ski prototype for curvature measurements on snow, (b) test the reliability and validity of the sensor-based w'' calculation by applying deformations that were similar to skiing in a dynamic setting, and (c) perform a proof-of-concept field measurement.

2. Materials and Methods

2.1. Further Developed PyzoFlex® Prototype for the Detection of the Local Ski Bending Curvature

An improved prototype was developed based on the PyzoFlex® ski demonstrator, which has already been presented for laboratory measurements [7,24]. PyzoFlex® sensors (www.pyzoflex.com (accessed on 1 July 2022), Joanneum Research Forschungsgesellschaft m.b.H, Franz-Pichler Str. 30, 8160 Weiz, Austria) are generally made from the ferroelectric copolymer P(VDF-TrFE) (poly(vinylidene-trifluoroethylene)) and are, therefore, inherently piezo- and pyro-electric [33]. For the sensor system that was used in this work, 24 sensor elements, divided into three foil segments (2×9 and 1×6 elements), were fabricated using screen printing on a single substrate. The segments were then cut out with a Trotec laser cutter and laminated to the surface of the ski (Atomic Redster G7; length: 1.82 m; radius: 19.6 m) with an adhesive tape from 3M™ series (VHB™ 5907F). Since the sensor foil is applied to the adhesive foil with the printed side facing downwards, this adhesive layer also acts as a seal against water penetration. Only the sensor flag at the edge of the foil is not adhered to the tape to enable the electrical connection to the readout electronics (Figure 1a).

For establishing contact between the sensor foils and the measurement data acquisition device (DAQ), the printed silver conductor lines were used with a foil connector from Amphenol (FCI Clincher™ Flex Connectors). To avoid short circuits at the silver contacts due to water penetration, the three contact points per ski were separately housed and sealed with silicone. All three sensor cables were merged at one point into the DAQ. In comparison to our former system [7], explicit emphasis was placed on the robustness of the system. Previously used ribbon cables were replaced by more robust sheathed and shielded multicore cables. One DAQ unit was installed per ski, which is capable of processing 16 of the 24 sensor channels in parallel. In addition, each evaluation unit has a so-called synchronization unit. On the one hand, this makes it possible to start and stop measurements, and on the other hand it enables synchronizing the data acquisition with the video recording by switching a LED that can be filmed by an external camera. The trigger time (i.e., the switching time of the LED) is stored in the database together with the sensor data and allows the synchronization with the video stream afterwards. Basically, the system can be operated in two modes. Either the data stream is transmitted instantly via Wi-Fi, displayed on a PC, and stored for further processing, or the data is stored on an on-board SD card and is read out and processed afterwards. Both the laboratory and field measurements were performed in the mode in which the data were recorded on the SD card. Figure 1b shows the complete PyzoFlex® demonstrator at the first system check on snow.

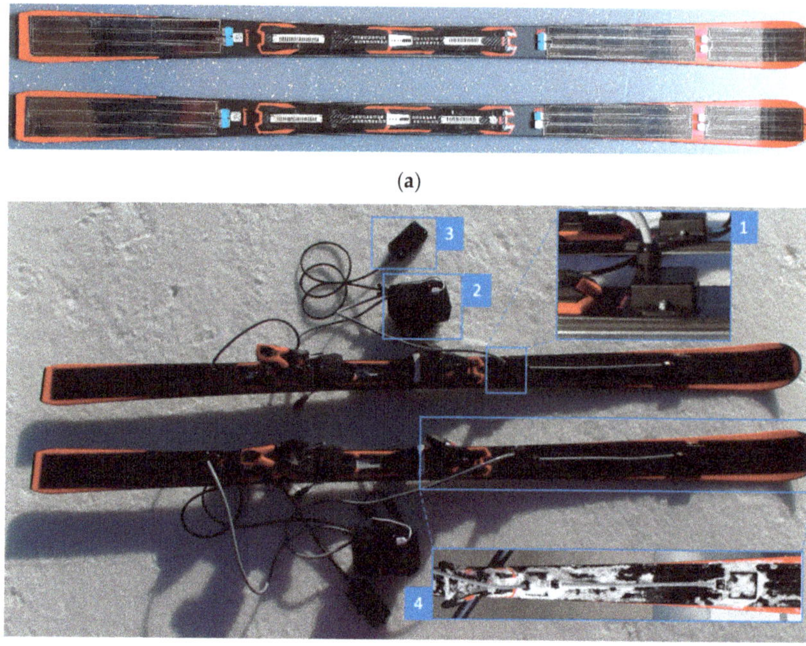

(**a**)

(**b**)

Figure 1. Ski prototype (**a**) after laminating the sensor foils and (**b**) after complete assimilation of all components. 1: Encased contact point that was sealed with silicone; 2: measurement data acquisition device (DAQ); 3: synchronization unit; 4: waterproof ski after system check.

In order to derive the local ski bending curvature from the measured charge response Q(t) of the sensor foils, our previously published calibration method was applied [7]. A charge amplifier circuit with a feedback capacity of $C_f = 20$ nF converts Q(t) into a corresponding output voltage $u_a(t)$. The mean curvature, $\overline{w''}$, is proportional to $u_a(t)$ and given by Equation (1):

$$\overline{w''}(u_a) = \frac{-C_f}{A \cdot v_{13} \cdot a \cdot P_r \cdot \zeta} \cdot (u_a - u_0) = (k \cdot u_a + d) \cdot f_d \quad (1)$$

where A is the interface area between the electrode and the ferroelectric material, v_{13} is the effective Poisson's ratio, a is a material constant close to unity, P_r is the remnant polarization resulting directly from the poling process, ζ denotes the distance to the neutral axis, and u_0 is the voltage baseline offset. Since ζ(x) is not available because it depends on the internal structure and the elastic properties of the rather complex ski construction, a calibration procedure is required for each combination of ski design, sensor design and alignment. Therefore, a two-point calibration was performed using a bending machine with a high-precision laser measurement system [7]. A linear regression model was established for each individual sensor and the calibration factors (slope k and the intercept d) were used to derive the segment's mean curvature $\overline{w''}$ from the voltage u_a. Furthermore, based on preliminary investigations, a dynamic correction factor (f_d = 0.78), which is constant for all the sensors, was included in Equation (1). This is necessary to compensate the relaxation effect that occurs in the quasi-static calibration process of the binding machine due to the viscoelastic behavior of the adhesive tape (Figure A3 in Appendix A). Moreover, during calibration the ski is not deformed from a neutral, but from a slightly pre-stressed initial position. The dynamic calibration factor also takes corrective action here.

2.2. *Reliability and Validity Assessment in the Laboratory*

2.2.1. Experimental Setup

An industrial articulated-arm robot (IRB 6400R, ABB AG, Brown-Boveri-Strasse 5, 8050 Zurich, Switzerland) was used to deform the PyzoFlex® technology-based ski prototype in a standardized dynamic setting. The robot's working space is thermally sealed off from the environment, which means that the temperature (T = 19 °C) in this area can be continuously adjusted. The robot's base frame, carousel, and weight-balancing system are located outside the thermally adjustable working space.

The ski was fixed to the robot arm by a test shoe. Underneath the ski was a frame that was made of aluminum profiles with rotatable synthetic rollers as supports at the front and the back of the ski. In the neutral position (NP), the ski rests unloaded on the synthetic rollers. The robot was programmed to cyclically deform the ski through five deformation modes. An overview of the deformation sequence that was described by the movement of the robot arm coordinate system (x_R, y_R, z_R) is shown in Figure 2. In the first mode, a central bending (CB) in the z_R-direction was performed, followed by a bending of the front (FB) and rear ski (BB) in y_R-z_R direction. Deformation modes 2 and 3 are intended to simulate forward and backward leaning of the skier. In modes 4 and 5, the ski is first edged to the left (EL) and then to the right (ER) by a lateral movement in the x_R-z_R-direction while tilting around the y_R axis. The edging and the resulting combined load of torsion and bending is supposed to simulate a carving turn. After ER, the cycle starts again from the beginning with CB. The ski returns to the NP between the modes, except for the transition from CB to FB. For the study, the entire deformation program (CB, FB, BB, EL, ER) was run 22 times at three different cycle times (CT_1 = 16 s, CT_2 = 12 s, CT_3 = 8.5 s), changing only the deformation velocity (v_{slow}, v_{medium}, v_{fast}) of the robot arm but not the deflection amplitudes. A PyzoFlex® prototype has already been tested for reliability on a pneumatic test bench over 90 cyclic oscillating repetitions [24]. For the reliability assessment, the first ten peak values of the sensor signal were compared with the middle and last ten peak values. Since a very reliable behavior is shown over 90 repetitions, the amount of 22 repetitions seems to be sufficiently large.

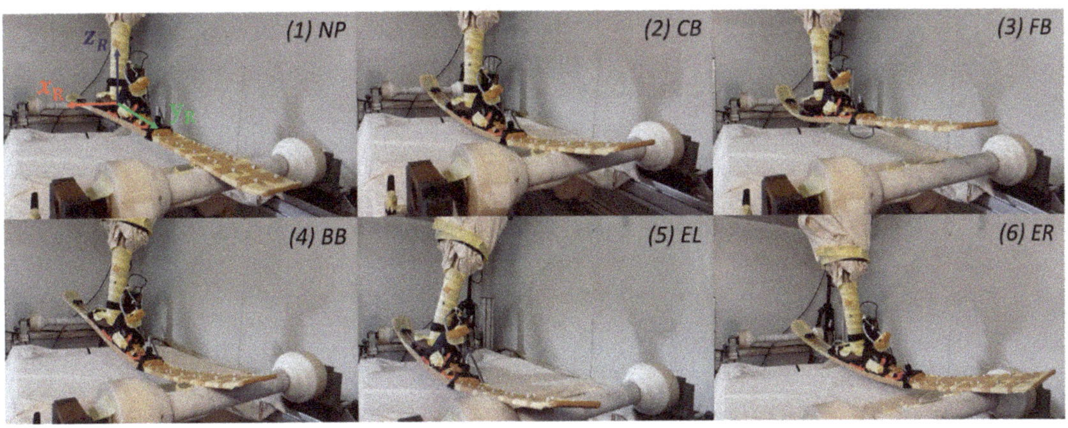

Figure 2. Experimental setup consisting of bending robot with the reference coordinate system (x_R, y_R, z_R) and the PyzoFlex® ski prototype that was equipped with the instrumented 3D markers. Deformation sequence of the ski on the bending robot: NP: neutral position; CB: central bending; FB: front bending; BB: back bending; EL: edging left; ER: edging right.

Reliability and validity assessments were performed with the instrumented ski that was described in Section 2.1, excluding twelve sensors (the front six sensors and the middle row). The corresponding sensors were not considered because, on the one hand, the number of sensor inputs of the DAQ was limited to 16 channels and, on the other hand, no comparison values from the gold standard measuring device were available for the validity assessment in the medial and anterior ski segments. To detect the three-dimensional ski deflection, an OMS (=gold standard) from Qualisys (Oqus 7+, Qualisys AB, Kvarnbergsgatan 2, 411 05 Göteborg, Sweden) with an acquisition frequency of 240 Hz was used. A total of eight active infrared cameras detected the markers within a capture volume of 3 m × 2 m × 1.5 m. The positions of the cameras and the description of the components can be found in a previously published Technical Note [33]. There were three plastic marker bases that were attached to each of the 12 sensors ($L_1, \ldots, L_6$ and $R_1, \ldots, R_6$) using an adhesive tape of the 3M™ VHB™ series (F9460PC) with a thickness of 0.05 mm. For this purpose, a 3D-printed template (Figure A1) was fabricated to ensure that the bases were applied as standardized as possible. In total, 36 infrared markers with a diameter of 6.5 mm and an internal M4 thread were screwed onto the plastic bases (Figure 3).

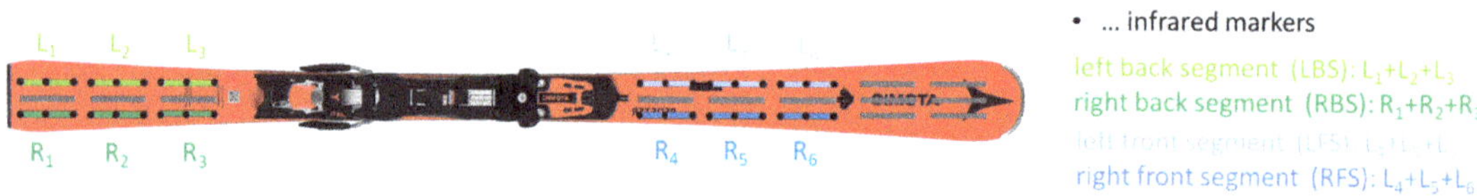

Figure 3. The PyzoFlex® ski prototype with the corresponding left ($L_1, \ldots, L_6$) and right ($R_1, \ldots, R_6$) sensor row. There were three infrared markers (black dots) that were instrumented per sensor, resulting in a total of 36 markers. The segments were grouped into posterior (LBS: left back segment, RBS: right back segment) and anterior deflection segments (LFS: left front segment, RFS: right front segment).

2.2.2. Data Processing–Motion Capture System

Based on the fact that no valid curvatures can be detected with Qualisys over a sensor with a length of 100 mm [33], the ski is divided into four bending segments (left rear segment (LBS), right rear segment (RBS), left front segment (LFS), and right front segment (RFS)) with a length of 340 mm (between two sensors is a gap of 20 mm). According to Figure 3, nine infrared markers are located in one bending segment. In principle, only three points are required to uniquely define w'' of a bending segment in three-dimensional space. However, to improve the signal-to-noise ratio and to minimize fluctuations in the curvature signal, a circle of the form:

$$C: \vec{x} = \vec{M} + R \cdot \cos(t) \cdot \vec{u} + R \cdot sin(t) \cdot \vec{v} \quad (2)$$

was fitted through nine measurement points using the least squares method [33]. $\vec{M}$ corresponds to the location vector of the center, R is the radius of the fitted circle, t refers to the free parameter of the parameter form and is between 0 and 2π, and the orthogonal unit vectors $\vec{u}$ and $\vec{v}$ lie in the plane of the circle. A customized algorithm in MATLAB (R2018B, MathWorks, Natick, MA, USA) was applied to perform the fitting routine based on Equation (2) and derive R. The mean curvature, $\overline{w''}$, can be calculated from the reciprocal of R using Equation (3):

$$\overline{w''} = \frac{1}{R} \quad (3)$$

Since the infrared markers were instrumented on the top side of the ski, but the calibration of the PyzoFlex® sensors was based on a high-precision laser measurement system (LK-H157, Keyence AG, Japan) on the bottom side, a discrepancy would arise when comparing the detected w'' from the sensor with the captured w'' from Qualisys. On the one hand, the w'' is geometrically larger on the upper side than on the lower side, and on the other hand, even slight differences in the instrumentation of the infrared markers can lead to a w'' offset in the NP. Therefore, the Qualisys system was calibrated preliminarily on a bending machine with a high-precision laser measurement system (Figure A2a) using Equation (4):

$$\overline{w_{bottom}''} = k_{laser} \cdot \overline{w''} + d_{laser} \tag{4}$$

There was a high linear correlation (r > 0.99) between the measured w'' from the laser and the detected w'' from the Qualisys system (Figure A2b). Equation (4) was, therefore, applied to calculate the segmental mean curvatures ($\overline{w_{LBS}''}, \overline{w_{RBS}''}, \overline{w_{LFS}''}, \overline{w_{RFS}''}$) at the bottom of the ski for the associated segments (LBS, RBS, LFS, RFS).

2.2.3. Data Processing—PyzoFlex® Sensor System

All PyzoFlex® data that were recorded during the measurement were filtered and linear drift-corrected using MATLAB. In particular, a 2nd order Butterworth high-pass filter with a cut-off frequency of 0.2 Hz was implemented to remove temperature-related signal fluctuations due to the pyroelectric effect from the raw data. Afterwards the sensor signals were calibrated as described in Section 2.1. The first and last cycles were not considered for the analysis based on the ringing artifacts by the filter in the border area, therefore 20 cycles remain for the analysis. For the validity assessment, segment-wise mean signals (LBS: (L_1 + L_2 + L_3)/3, RBS: (R_1 + R_2 + R_3)/3, LFS: (L_4 + L_5 + L_6)/3 and RFS: (R_4 + R_5 + R_6)/3) are formed, since a valid curvature can only be detected over a length of three sensors [33]. Since the maximum values of the sensor signals are decisive for the data processing, these values were determined by a custom peak detection algorithm in MATLAB.

2.2.4. Statistical Analysis

To determine the dynamic performance of PyzoFlex® technology-based sensors, the aspects of instrument reliability (within the PyzoFlex® sensor system) and criterion validity (between systems) were examined. For the statistical analysis, 12 sensors (L_1 to L_6 and R_1 to R_6) with the corresponding peak values (in m^{-1}) at CB, FB, BB, EL, and ER were considered.

To assess the reliability, data were examined statistically (Shapiro–Wilk test) for normal distribution. Changes in the mean (CIM) for three dynamic situations expressed by the deformation velocity (v_{slow}, v_{medium}, v_{fast}) were analyzed using a paired samples *t*-test (level of significance $p < 0.05$). For the assessment of relative test-retest reliability, the typical error of measurement represented as the coefficient of variation (CV) and the intraclass correlation coefficient (ICC 3.1) were utilized. ICC 3.1 was interpreted as unacceptable (≤0.74), good (0.75–0.89), and excellent (≥0.9) [34]. Successive pairs of trials (cycle 1 with cycle 10, cycle 10 with cycle 20) were considered for CIM, CV, and ICC 3.1. In addition, an analogous test-retest reliability between different deformation velocities (v_{slow} vs. v_{medium} and v_{slow} vs. v_{fast}) was analyzed by considering the peak values of cycle 1, 10, and 20.

Pearson's correlation coefficient was used for criterion validity between PyzoFlex® and the Qualisys system. The magnitude of correlations was graded as follows: r < 0.45, impractical; 0.45–0.70, very poor; 0.70–0.85, poor; 0.85–0.95, good; 0.95–0.995, very good; and >0.995, excellent [35]. To describe the agreement between the sensor system and the Qualisys measurement method, a Bland–Altman plot was employed [36]. The accuracy, expressed by the systematic bias, and the precision, expressed by the standard derivation (SD), were evaluated to assess the limits of agreement (LoA) (±0.96 × SD). All the metrics were calculated using MATLAB and a custom spreadsheet [35], except for the paired samples *t*-test statistics, for this IBM SPSS Statistics V.26.0 (SPSS Inc., Chicago, IL, USA) was used.

2.3. *Proof-of-Concept Field Measurement*

The proof-of-concept field measurement was carried out on 2 March 2021 in the Petersbründel ski area (Austria). The experimental setup is illustrated in Figure 4. The test person (A-level ski instructor) performed 24 carving turns (12 left turns and 12 right turns) with long radii within a defined measurement corridor. A follow cam and a body cam were used to determine the turn switch points (TSPs) of the carving turns. At the beginning and end of the run, the self-built synchronization unit was used to perform the sync event to synchronize the PyzoFlex® data with the video data.

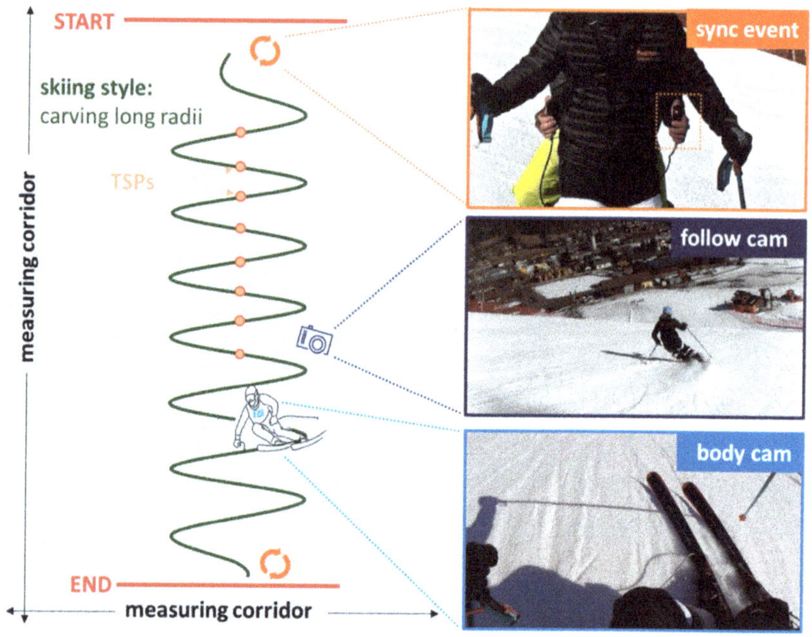

Figure 4. Experimental field setup. TSPs: Turn switch points.

Only the left row of the right ski was used for the descriptive analysis. The recorded sensor signals were filtered and linearly drift-corrected analogously to the laboratory setting and additionally processed with a custom zero-update algorithm using MATLAB. Further, the data were time normalized over a left-right turn combination and then the mean w'' ± standard error (SE) was calculated.

3. Results

3.1. Descriptive Report of Lab Data

Figure 5 shows the detected w'' as a result of the sequence of five deformation modes (CB, FB, BB, EL, and ER) at v_{fast}. For the illustration of the 12 sensors, the mean value ± SD (in m^{-1}) over 20 repetitions was calculated for each measuring point. A differentiation is made between the left (L_1 to L_6) and right sensor rows (R_1 to R_6), showing all the rear sensors in blue (L_1, L_2, L_3; R_1, R_2, R_3) and the front sensors in green (L_4, L_5, L_6; R_4, R_5, R_6). The five deformation modes result in four peaks per sensor, since only the front or rear ski were deformed in the case of FB or BB. The sensors in the rear segments 2 and 3 are activated the most, followed by the sensors in segments 6, 5, and 4. The sensors in the first segment are partially above the support and are the least curved.

3.2. Instrument Reliability and Accuracy

Table 1 shows test-retest reliability results for both repetition pairs (cycle 1 vs. cycle 10 and cycle 10 vs. cycle 20) at three different deformation velocities (v_{slow}, v_{medium}, and v_{fast}). The CIM over all the comparisons ranged between −0.69% and 0.24%. The maximum CV value of 0.28% can be noted when comparing cycle 1 with 10 at v_{medium}. ICC 3.1 was very close to 1 (ICC 3.1 > 0.999, $p < 0.001$) for all cases and significant differences are detectable in four of the six comparisons.

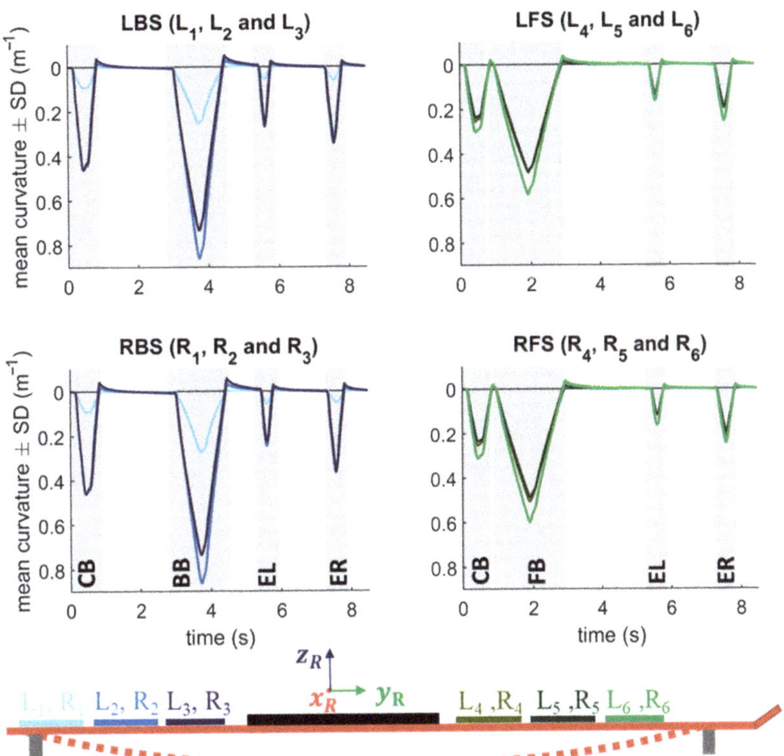

Figure 5. The PyzoFlex® sensor signals (mean +/− standard deviation (SD)) over 20 cycles during dynamic loading with v_{fast}. The rear sensors (R_1, R_2, R_3, L_1, L_2, and L_3) are shown in blue and the front sensors (R_4, R_5, R_6, L_4, L_5, and L_6) in green. The ski was deformed with a bending robot through five deformation modes (CB: central bend; FB: front bend; BB: rear bend; EL: edge left; ER: edge right).

Table 1. Analysis of instrument reliability from the sensor system.

		p (*t*-Test)	CIM (%, 95% CI)	CV (%, 95% CI)	ICC 3.1 (95% CI)
v_{slow}	cycle 1 vs. 10	0.010	0.09 (0.00–0.18)	0.26 (0.22–0.31)	0.999 (0.999–0.999)
	cycle 10 vs. 20	n.s.	0.01 (−0.06–0.08)	0.21 (0.18–0.26)	0.999 (0.999–0.999)
v_{medium}	cycle 1 vs. 10	0.016	0.14 (0.05–0.24)	0.28 (0.24–0.34)	0.999 (0.999–0.999)
	cycle 10 vs. 20	n.s.	−0.01 (-0.08–0.07)	0.22 (0.19–0.26)	0.999 (0.999–0.999)
v_{fast}	cycle 1 vs. 10	<0.001	0.24 (0.17–0.31)	0.21 (0.18–0.25)	0.999 (0.999–0.999)
	cycle 10 vs. 20	<0.001	−0.69 (-0.75–0.63)	0.17 (0.15–0.21)	0.999 (0.999–0.999)

CIM: change in mean; CI: confidence interval; CV: coefficient of variance; ICC 3.1: intraclass correlation coefficient.

Table 2 shows the test-retest reliability results between different velocities (v_{slow} vs. v_{medium} and v_{slow} vs. v_{fast}). The CIM is greater at a higher velocity difference (3.03%) than when comparing v_{slow} with v_{medium} (1.85%). The maximum CV value is 3.05% and the ICC 3.1 is greater than 0.997. A significant difference can be found in both comparisons.

Table 2. Analysis of the reliability of the instruments of the sensor system in respect of different deformation velocities.

		p (t-Test)	CIM (%, 95% CI)	CV (%, 95% CI)	ICC 3.1 (95% CI)
Cycle 1, 10 and 20	v_{slow} vs. v_{medium}	<0.001	1.84 (1.25–2.43)	3.05 (2.74–3.38)	0.997 (0.996–0.998)
	v_{slow} vs. v_{fast}	<0.001	3.03 (2.61–3.46)	2.18 (1.99–2.41)	0.999 (0.998–0.999)

CIM: change in mean; CI: confidence interval; CV: coefficient of variance; ICC 3.1: intraclass correlation coefficient.

3.3. Criterion Validity

Figure 6 shows the signal of the PyzoFlex® ski in red and the gold standard system (Qualisys) in blue over three bending cycles at v_{medium}. In the BB situation, the biggest difference between Qualisys and PyzoFlex is obviously seen in the peaks of LBS and RBS. The Qualisys data in the grey marked area cannot be judged, as valid results from Qualisys can only be expected from w'' of 0.13 m^{-1} upwards [33].

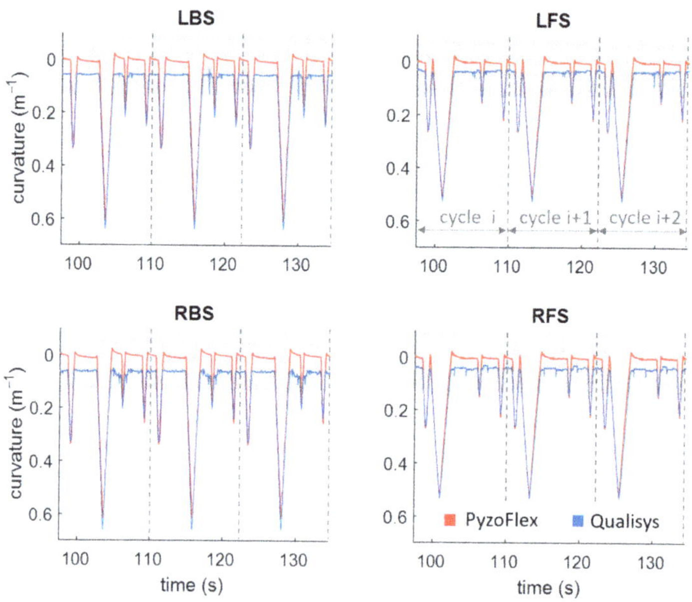

Figure 6. PyzoFlex® signal in red and Qualisys (QTM) signal in blue over three cycles during dynamic loading at v_{medium}. A differentiation is made between the posterior (LBS: left posterior segment, RBS: right posterior segment) and the anterior deflection segments (LFS: left anterior segment, RFS: right anterior segment).

In Figure 7a, data from the PyzoFlex® sensor system were correlated to the criterion instrument (Qualisys system) to determine validity. The slope of the function is 1.03 and the coefficient of determination R^2 is 0.97. Figure 7b shows the Bland–Altman plot with the systematic bias of 0.003 m^{-1} and the LoA (−0.096 m^{-1} and 0.090 m^{-1}).

3.4. Proof-of-Concept Field Measurement

Figure 8 shows the time-normalized w'' (mean ± SE) of 24 carving turns with long radii. The TSP from a right to a left turn is at 0% and 100% and from a left to right turn at 50%. The rear sensors are shown in blue and the front sensors in green. The sensor behind the binding plate (L_3) shows the highest peak with a value of about 0.8 m^{-1}.

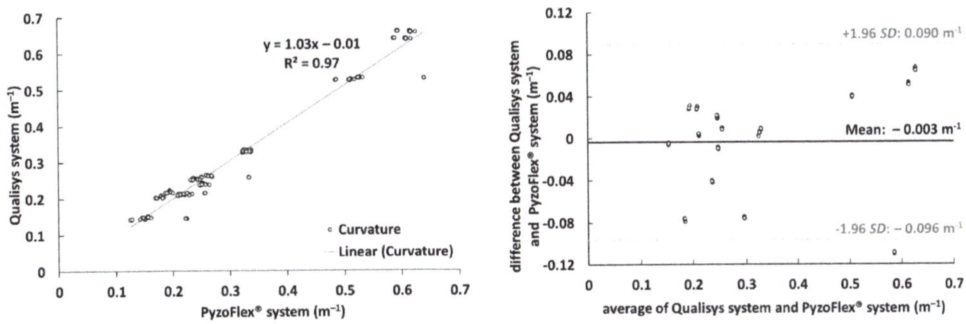

(**a**) (**b**)

Figure 7. (**a**): Correlation between the curvature (m^{-1}) that was measured by Qualisys system (criterion instrument) and PyzoFlex® sensor system. (**b**): Bland–Altman plot showing the difference against the average of Qualisys system and PyzoFlex® sensor system with limits of agreement (LoA) (dotted lines). SD: Standard deviation.

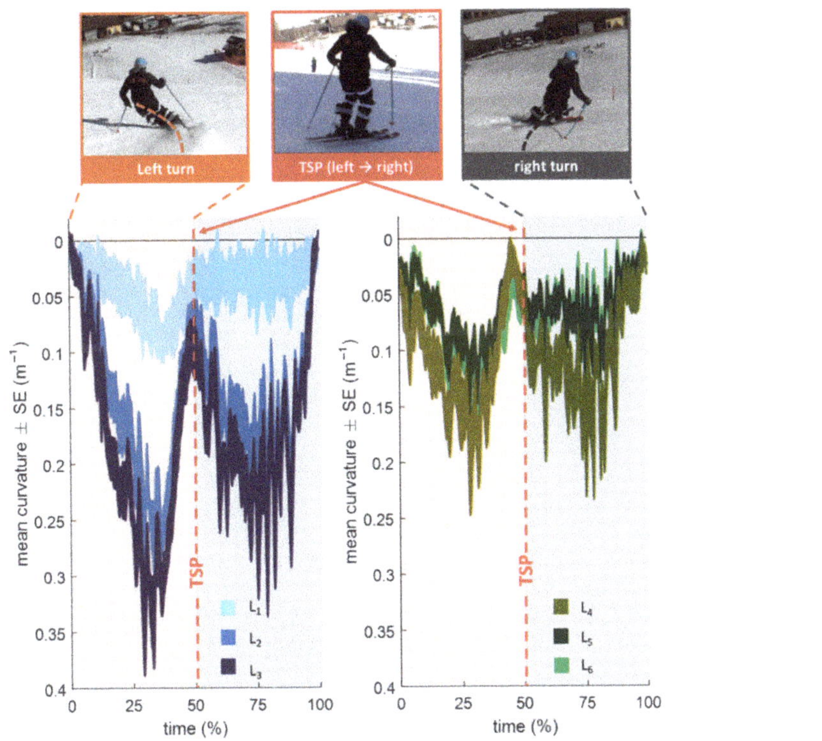

Figure 8. Result of the proof-of-concept field measurement with carving long radii (12 left and 12 right turns). The plot differentiates between the left and right turns. The rear sensors (L_1, L_2, L_3) are shown in blue and the front sensors (L_4, L_5 and L_6) in green tones. SE: standard error.

4. Discussion

The objective of this study was to present an improved PyzoFlex® ski prototype for the detection of w'', to evaluate its test-retest reliability and criterion validity in a dynamic laboratory setting, and to perform a proof-of-concept field measurement. The PyzoFlex®

prototype can be considered as an accurate and precise instrument, as the data showed a very good agreement with the reference system (Qualisys) and produced plausible w'' characteristic on the snow.

The results have shown that the ski prototype provides highly reliable data. Within the velocities, the max CIM was 0.69%, the max CV was 0.28%, and the ICC 3.1 was found to be consistently close to 1 (all ICC 3.1 > 0.999), indicating excellent reliability (Table 1) [34]. Comparing the values between the velocities, both the CIM and CV are slightly higher in the range of 1.84% to 3.05%, with the ICC 3.1 again very close to 1 (Table 2). Out of the total eight comparisons, there were six significant differences, with the largest difference evident for v_{slow} vs. v_{fast} (CIM = 3.03%, $p < 0.001$). When a comparable reliability test was carried out with the Qualisys data for v_{slow} vs. v_{fast}, a small but significant difference (CIM = −0.14%, $p < 0.001$) is also noticeable. However, the small but significant difference argues for the accuracy of the PyzoFlex® system, as the differences can be at least partially explained by the Qualisys data [7].

In terms of validity, there was also good agreement between the calculated w'' of the sensor signal and the captured w'' by the Qualisys system. The magnitude of the linear relationship between the two systems, expressed by the correlation coefficient of r = 0.985, is very good (Figure 7a) [35]. Moreover, a high accuracy was shown since the systematic bias was only $-0.003\ m^{-1}$. The precision of the signal denoted by the SD ($0.047\ m^{-1}$) and LoA ($-0.096\ m^{-1}$ and $0.090\ m^{-1}$) is consistent with the other results and indicates high validity (Figure 7b). However, it is evident that the accuracy for the LBS and RBS in the BB situation is comparatively low (Figure 6). This could be since the posterior segments have the greatest w'' and thus the relaxation effect of the tape is more pronounced, making the PyzoFlex® w'' appear to be smaller. As can be seen in Figure 6, there is hardly any difference between the parallel mounted sensors. In particular, the differences in EL and ER were expected, as this represents a combined load of torsion and bending, and this should be evident in the differential signal of the parallel mounted sensors. However, this result is consistent with the finding that ski deflection consists mainly of bending and that torsional deformation has only a minor influence on the shape of the running surface [18].

In previous studies, little attention was paid to ski w'' except for our two studies with the PyzoFlex® prototype [7,24]. First, the reliability of the sensors in the presented experimental setup is comparable to those in the quasi-static setting, as the CIM is −1.41%, the max CV is 1.45%, and the ICC 3.1 is >0.961 [7]. Second, the reliability assessments are also comparable to the results of a oscillating bending measurement (CIM = 0.20%, CV = 0.63% and the ICC 3.1 > 0.997) [24]. However, most studies have focused on sensor-based detection of the deflection line $w(x)$, but not on the analysis of $w''(x)$, and further, no reliability and validity tests have been performed, so that no comparison with previously used systems and prototypes is possible.

The proof-of-concept measurement has shown on the one hand that the prototype is stable, robust, and waterproof and on the other hand that plausible w'' values can be measured. The w'' shows a characteristic pattern with the highest w'' between the TSPs (Figure 8). However, the rapidly fluctuating w'' within a turn is consistent with the findings that the ski shape changes rapidly within a short time period [18]. When comparing the front and rear segments, temporal differences can be observed, as the front sensors are activated at an earlier stage relative to the TSPs. Based on the analysis of the video data, this seems plausible since during TSPs often the rear end of the ski is in the air and the front end is pre-stressed (see middle image in Figure 8). According to the definition of a carving turn, that the rear section of the ski glides in the front groove that is formed by the front section [5], an earlier temporal activation of the front sensors seems to be reasonable. The rear segment is more deformed, which is consistent with both the measurements with the bending robot that was described above and the results of the prototypes that are based on strain gauges [16,17]. Furthermore, it is also reported that the effective bending radius of the ski in carving turns is about 10 m, while the true turn radius is 20 m [18]. This can also be confirmed in this work, as the max value of the mean w'' across all sensors

$(L_1 + , \ldots , + L_6)/6)$ is 0.19 m^{-1}. This is equivalent to a radius of about 5.3 m, which is at least twice as large as the true turn radii.

The study that is presented has some limitations. First, single sensors with a length of 100 mm cannot be checked for validity, as this is outside the accuracy range of the 3D motion capture system (Qualisys) [33]. Averaging over three sensors is, therefore, a necessary procedure for the validity assessment. Since there is a distance of 2 cm between the sensors, two small sub-segments are not taken into account for w'' detection. Secondly, it was observed that the sensors that were calibrated in a quasi-static environment would not give valid results in a dynamic setting, so a correction factor (f_d) was included in the curvature model. This is primarily justified by the relaxation phenomena due to the viscoelastic behavior of the tape (Figure A3). Additional measurements have shown that, on the one hand, the relaxation effect is lower with thinner adhesive tapes (200 µm instead of the 1.1 mm version of the tape) and, on the other hand, larger sensor deflections can be observed due to the more direct connection. For the future, dynamic calibration appears to be a reasonable option. Therefore, it seems appropriate to develop a test bench with high-precision laser measurement systems on which the ski can be dynamically deformed and the single sensors calibrated. This would also have the advantage that, in contrast to Qualisys, w'' below 0.13 m^{-1} can also be recorded. Furthermore, it should be noted that due to the pyroelectric nature of the sensors, the temperature influence has not yet been sufficiently clarified. Basically, it is assumed that the low-frequency temperature changes can be filtered out with a high-pass filter. Measurements in a cold chamber are necessary to clearly determine the temperature-related influence.

The novel application needs to be progressively developed in different areas. In the current prototype, three sensor elements per ski communicate with the DAQ via two interfaces. In perspective, one large sensor element is to be used over the entire length of the ski so that the interfaces are reduced to one connection behind the binding plate. This has the advantage that the conductor paths run under the binding plate and the sensors can also be integrated in this segment. For easier handling, the Raspberry Pi-based DAQ should be replaced by a microcontroller-based system and screwed directly to the ski boot. Perspectively, an integration of the sensor foils into the ski construction would be advised to protect the sensors from mechanical damage and environmental influences.

Further in-depth snow measurements are required to investigate the equipment- and technique-specific ski-snow interactions. The knowledge that is gained can flow into various development and application fields. First, the ski deflection and behavior that was identified on the basis of the sensor activity, might serve as a potential application for injury analysis and prevention. In terms of product fitting, the PyzoFlex® technology-based ski could help to find the appropriate ski, not only in terms of safety but also in terms of performance. For example, if the deflection of the ski is too low according to the measured w'', it seems appropriate to choose a ski with a lower bending stiffness. On the other hand, personalized feedback can be provided to improve the skiing technique and performance. Of particular interest is also the extent to which the segmental and temporal curvature characteristics differ depending on the performance level (high-performance athletes vs. amateur athletes) and the technique that is used (carving, parallel ski steering, snowplough steering, etc.). In order to shed light on the construct "ski deflection", it seems reasonable to use additional measurement systems for snow investigations. Thus, plausibility checks, pattern recognition, and technical validation can be carried out. In terms of discriminant validity, correlations between different measures (such as edge angle or radial force) can provide revealing information. Furthermore, the indoor test methodology for ski development is to be designed more realistically based on the field findings in order to increase the safety and stability of product engineering.

5. Conclusions

Ski deflection is the result of a multifactorial interaction between the ski and the snow and has a crucial influence on skiing performance and on the quality of a turn. Up to now, there are no commercial applications that measure ski deflection during alpine skiing. The ski prototype that was presented here, based on PyzoFlex® technology, opens a potential field of application for smart ski equipment, since segmental w'' can be reliably and validly detected. In addition to the possibility of a real-time feedback system, application in the field of product fitting, and injury analysis, conclusions can also be drawn about indoor testing methodology for ski development. For the future, however, further improvements of the prototype are planned (e.g., the integration of the sensor foils into the ski, enhancement and downsizing of the DAQ and implementation on the ski boot, reduction of the number of interfaces on the ski).

Author Contributions: Conceptualization, C.T., J.K. and T.S.; methodology, C.T., J.K., A.T. and W.T.; validation, J.K. and T.S.; formal analysis, T.S.; investigation, C.T., A.T. and W.T.; resources, A.T. and H.H.; data curation C.T.; writing—original draft preparation, C.T.; writing—review and editing, J.K. and T.S.; visualization, C.T.; supervision, J.K. and T.S.; project administration, T.S.; funding acquisition, J.K. and T.S. All authors have read and agreed to the published version of the manuscript.

Funding: This work was partly funded by the Austrian Federal Ministry for Transport, Innovation and Technology, the Austrian Federal Ministry for Digital and Economic Affairs, and the federal state of Salzburg, as well as from the European Union's Horizon 2020 research and innovation program.

Institutional Review Board Statement: Not applicable.

Informed Consent Statement: Not applicable.

Data Availability Statement: The data presented in this study are available on request from the corresponding author.

Acknowledgments: The authors would like to thank Rupert Oberreiter for his support during the laboratory measurements. We would also like to thank Sascha Posch and Michael Lasshofer for their participation in the field tests and Philipp Schäffner for his support in the review process.

Conflicts of Interest: The authors declare no conflict of interest.

Appendix A

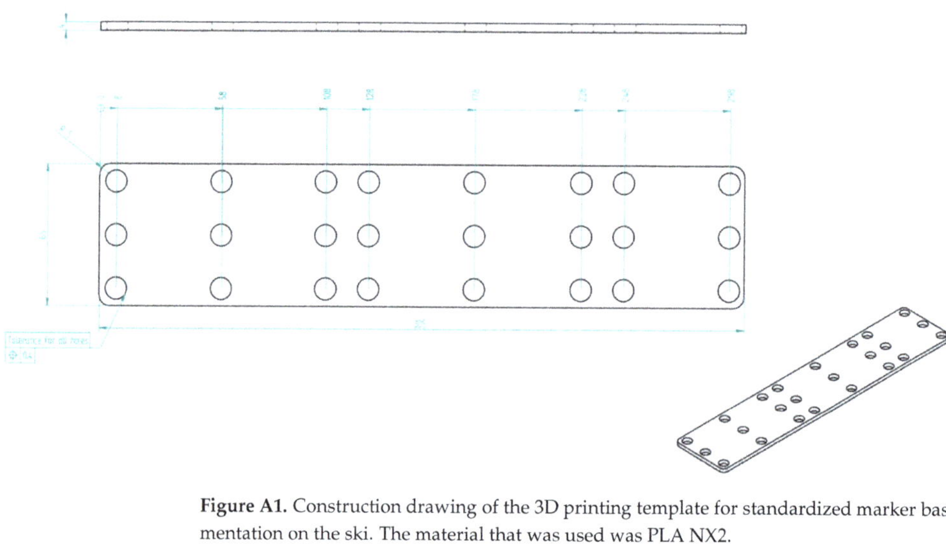

Figure A1. Construction drawing of the 3D printing template for standardized marker bases instrumentation on the ski. The material that was used was PLA NX2.

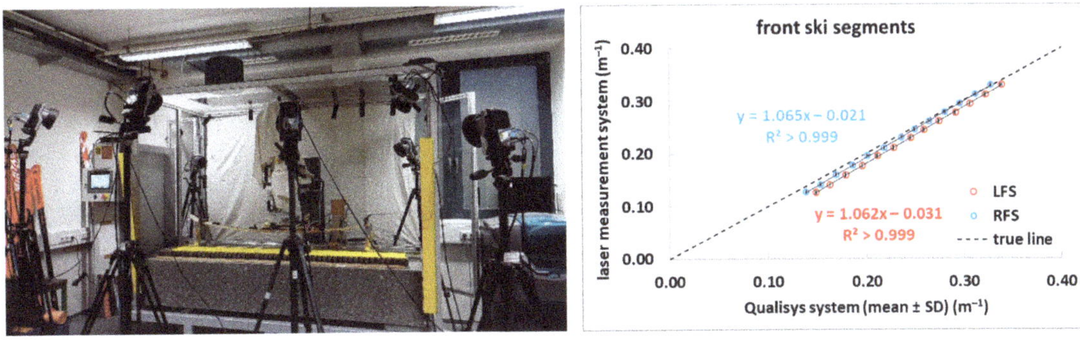

(**a**) (**b**)

Figure A2. (**a**): The ski was deformed on a bending machine at 13 different load levels (100 N, 120 N, . . . , 320 N, 340 N). (**b**): Correlation of the Qualisys system with a high precision laser measurement system (LK-H157, Keyence AG, Japan). The calibration was performed using the corresponding compensation lines in the form $\overline{w_{bottom}''} = k_{laser} \cdot \overline{w''} + d_{laser}$. LFS: left front segment; RFS: right front segment; SD: standard deviation.

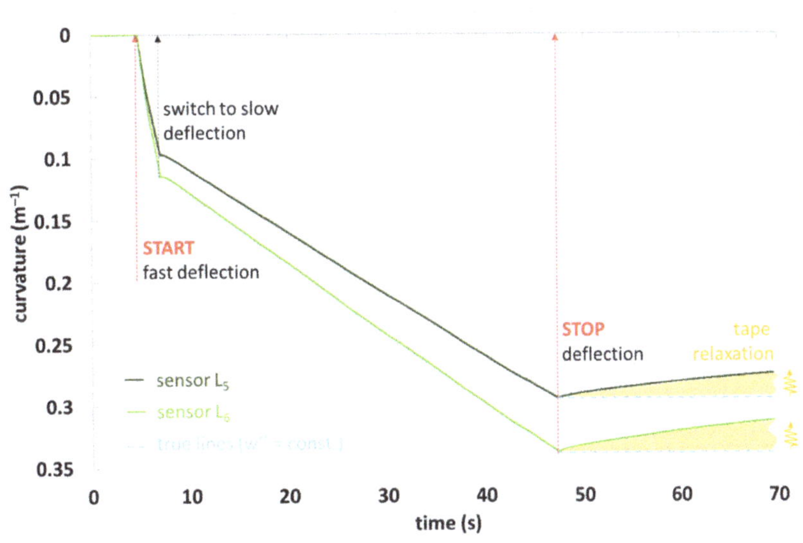

Figure A3. Temporal change of the signal from sensor L_4 and L_5 during the quasi-static deformation on the bending machine. At the end of the deflection, the relaxation process of the sensor signals due to the viscoelastic behavior of the tape is visible.

References

1. Lieu, D.; Mote, C. Mechanics of the turning snow ski. In *Skiing Trauma and Safety: Fifth International Symposium*; Johnson, R.J., Mote, C.D., Eds.; ASTM International: Philadelphia, PA, USA, 1985; Volume 1, pp. 117–140.
2. Renshaw, A.A.; Mote, C., Jr. A model for the turning snow ski. *Int. J. Mech. Sci.* **1989**, *31*, 721–736. [CrossRef]
3. Reid, R.C.; Haugen, P.; Gilgien, M.; Kipp, R.W.; Smith, G.A. Alpine Ski Motion Characteristics in Slalom. *Front. Sports Act. Living* **2020**, *2*, 25. [CrossRef] [PubMed]
4. Brown, C.; Outwater, J.O. On the skiability of snow. In *Skiing Trauma and Safety: Seventh International Symposium, ASTM STP 1022*; Johnson, R., Mote, C., Binet, M., Eds.; ASTM International: Philadelphia, PA, USA, 1989; Volume 1, pp. 329–336.
5. Federolf, P.; Roos, M.; Lüthi, A.; Dual, J. Finite element simulation of the ski–snow interaction of an alpine ski in a carved turn. *Sports Eng.* **2010**, *12*, 123–133. [CrossRef]

6. Adelsberger, R.; Aufdenblatten, S.; Gilgien, M.; Tröster, G. On bending characteristics of skis in use. *Procedia Eng.* **2014**, *72*, 362–367. [CrossRef]
7. Thorwartl, C.; Kröll, J.; Tschepp, A.; Schäffner, P.; Holzer, H.; Stöggl, T. A novel sensor foil to measure ski deflections: Development and validation of a curvature model. *Sensors* **2021**, *21*, 4848. [CrossRef]
8. Nachbauer, W.; Rainer, F.; Schindelwig, K. Effects of ski stiffness on ski performance. In Proceedings of the Engineering of Sport 5, Winfield, IL, USA, 13–16 September 2004; pp. 472–478.
9. Rainer, F.; Nachbauer, W.; Schindelwig, K.; Kaps, P. On the measurement of the stiffness of skis. In *Science and Skiing III*; Müller, E., Bacharach, D., Klika, R., Eds.; Meyer & Meyer Sport: Aachen, Germany, 2005; Volume 1, pp. 136–147.
10. Deak, A.; Jorgensen, J.; Vagners, J. The engineering characteristics of snow skis—part 1: Static bending and torsional characteristics. *J. Eng. Ind.* **1975**, *97*, 131–137. [CrossRef]
11. Lüthi, A.; Federolf, P.; Fauve, M.; Rhyner, H. Effect of bindings and plates on ski mechanical properties and carving performance. In *The Engineering of Sport 6*; Hubbard, M., Metha, R.D., Pallis, J.M., Eds.; Springer: Winfield, IL, USA, 2006; Volume 1, pp. 299–304.
12. De Gobbi, M.; Petrone, N. Structural behaviour of slalom skis in bending and torsion (P269). In *The Engineering of Sport 7*; Estivalet, M., Brisson, P., Eds.; Springer: Paris, France, 2008; Volume 2, pp. 643–652.
13. Truong, J.; Brousseau, C.; Desbiens, A.L. A Method for Measuring the Bending and Torsional Stiffness Distributions of Alpine Skis. *Procedia Eng.* **2016**, *147*, 394–400. [CrossRef]
14. Clifton, P.; Subic, A.; Mouritz, A. Snowboard stiffness prediction model for any composite sandwich construction. *Procedia Eng.* **2010**, *2*, 3163–3169. [CrossRef]
15. Kagawa, H.; Yoneyama, T.; Tatsuno, D.; Scott, N.; Osada, K. Development of a Measuring System on Ski Deflection and Contacting Snow Pressure in Turns. In *Science and Skiing IV*; Müller, E., Lindinger, S., Stöggl, T., Eds.; Meyer & Meyer Sport: Aachen, Germany, 2009; Volume 1, pp. 281–291.
16. Yoneyama, T.; Kitade, M.; Osada, K. Investigation on the ski-snow interaction in a carved turn based on the actual measurement. *Procedia Eng.* **2010**, *2*, 2901–2906. [CrossRef]
17. Tatsuno, D.; Yoneyama, T.; Kagawa, H.; Scott, N.; Osada, K. Measurement of Ski Deflection and Ski-Snow Contacting Pressure in an Actual Ski Turn on the Snow Surface. In *Science and Skiing IV*; Müller, E., Lindinger, S., Stöggl, T., Eds.; Meyer & Meyer Sport: Aachen, Germany, 2009; Volume 1, pp. 505–515.
18. Yoneyama, T.; Scott, N.; Kagawa, H.; Osada, K. Ski deflection measurement during skiing and estimation of ski direction and edge angle. *Sports Eng.* **2008**, *11*, 3–13. [CrossRef]
19. Schindelwig, K.; Hellberg, F.; Nachbauer, W.; Schretter, H. Measuring dynamic ski behavior with strain gauges. In Proceedings of the XXIV International Symposium on Biomechanics in Sports, Salzburg, Austria, 14–18 July 2006.
20. Fauve, M.; Auer, M.; Lüthi, A.; Rhyner, H.; Meier, J. Measurement of Dynamical Ski Behavior during Alpine Skiing. In *Science and Skiing IV*; Müller, E., Lindinger, S., Stöggl, T., Eds.; Meyer & Meyer Sport: Aachen, Germany, 2009; Volume 1, pp. 195–206.
21. Kos, A.; Umek, A. Smart sport equipment: SmartSki prototype for biofeedback applications in skiing. *Pers. Ubiquitous Comput.* **2018**, *22*, 535–544. [CrossRef]
22. Umek, A.; Kos, A. Smart equipment design challenges for real time feedback support in sport. *Facta Universitatis. Ser. Mech. Eng.* **2018**, *16*, 389–403. [CrossRef]
23. Yoneyama, T.; Kagawa, H.; Tatsuno, D.; Kitade, M.; Osada, K.; Shigehara, S. Effect of flexural stiffness distribution of a ski on the ski–snow contact pressure in a carved turn. *Sports Eng.* **2021**, *24*, 2. [CrossRef]
24. Thorwartl, C.; Holzer, H.; Kröll, J.; Tschepp, A.; Stöggl, T. Sensor-Based Curvature Detection of Alpine Skis. In Proceedings of the 24th International Congress on Snow Sport Trauma and Safety—37th Congress of the International Society for Snowsports Medicine, Mônetier-les-Bains, France, 21–27 March 2021; p. 58.
25. Corazza, S.; Mündermann, L.; Gambaretto, E.; Ferrigno, G.; Andriacchi, T.P. Markerless motion capture through visual hull, articulated icp and subject specific model generation. *Int. J. Comput. Vis.* **2010**, *87*, 156. [CrossRef]
26. van der Kruk, E.; Reijne, M.M. Accuracy of human motion capture systems for sport applications; state-of-the-art review. *Eur. J. Sport Sci.* **2018**, *18*, 806–819. [CrossRef]
27. Windolf, M.; Götzen, N.; Morlock, M. Systematic accuracy and precision analysis of video motion capturing systems—exemplified on the Vicon-460 system. *J. Biomech.* **2008**, *41*, 2776–2780. [CrossRef]
28. Monnet, T.; Samson, M.; Bernard, A.; David, L.; Lacouture, P. Measurement of three-dimensional hand kinematics during swimming with a motion capture system: A feasibility study. *Sports Eng.* **2014**, *17*, 171–181. [CrossRef]
29. Gilgien, M.; Spörri, J.; Limpach, P.; Geiger, A.; Müller, E. The effect of different global navigation satellite system methods on positioning accuracy in elite alpine skiing. *Sensors* **2014**, *14*, 18433–18453. [CrossRef]
30. Van der Kruk, E. Modelling and Measuring 3D Movements of a Speed Skater. Master's Thesis, TU Delft, Delft, The Netherlands, 2013.
31. Maletsky, L.P.; Sun, J.; Morton, N.A. Accuracy of an optical active-marker system to track the relative motion of rigid bodies. *J. Biomech.* **2007**, *40*, 682–685. [CrossRef]
32. Liu, H.; Holt, C.; Evans, S. Accuracy and repeatability of an optical motion analysis system for measuring small deformations of biological tissues. *J. Biomech.* **2007**, *40*, 210–214. [CrossRef] [PubMed]
33. Thorwartl, C.; Stöggl, T.; Teufl, W.; Holzer, H.; Kröll, J. Curvature Detection with an Optoelectronic Measurement System Using a Self-Made Calibration Profile. *Sensors* **2022**, *22*, 51. [CrossRef] [PubMed]

34. Koo, T.K.; Li, M.Y. A guideline of selecting and reporting intraclass correlation coefficients for reliability research. *J. Chiropr. Med.* **2016**, *15*, 155–163. [CrossRef] [PubMed]
35. Hopkins, W.G. Spreadsheets for analysis of validity and reliability. *Sportscience* **2017**, *21*, 36–44.
36. Bland, J.M.; Altman, D.G. Measuring agreement in method comparison studies. *Stat. Methods Med. Res.* **1999**, *8*, 135–160. [CrossRef] [PubMed]

sensors

MDPI

Article

Detecting and Visualizing Stops in Dance Training by Neural Network Based on Velocity and Acceleration

Yuuki Jin, Genki Suzuki * and Hiroyuki Shioya

Division of Information and Electronic Engineering, Muroran Institute of Technology, Muroran 050-8585, Japan; kokoronomotono@gmail.com (Y.J.); shioya@mmm.muroran-it.ac.jp (H.S.)
* Correspondence: suzuki@mmm.muroran-it.ac.jp

Abstract: Various genres of dance, such as Yosakoi Soran, have contributed to the health of many people and contributed to their sense of belonging to a community. However, due to the effects of COVID-19, various face-to-face activities have been restricted and group dance practice has become difficult. Hence, there is a need to facilitate remote dance practice. In this paper, we propose a system for detecting and visualizing the very important dance motions known as stops. We measure dance movements by motion capture and calculate the features of each movement based on velocity and acceleration. Using a neural network to learn motion features, the system detects stops and visualizes them using a human-like 3D model. In an experiment using dance data, the proposed method obtained highly accurate stop detection results and demonstrated its effectiveness as an information and communication technology support for remote group dance practice.

Keywords: neural network; motion sensor; dance analysis; data visualization

Citation: Jin, Y.; Suzuki, G.; Shioya, H. Detecting and Visualizing Stops in Dance Training by Neural Network Based on Velocity and Acceleration. *Sensors* **2022**, *22*, 5402. https://doi.org/10.3390/s22145402

Academic Editor: Vesa Linnamo

Received: 23 June 2022
Accepted: 15 July 2022
Published: 20 July 2022

1. Introduction

There are various genres of dance around the world, such as ballet, hip hop, and Japanese dance. Like sports, dance is popular for its strong exercise benefit [1–7]. Dance also has artistic aspects linked to the culture of a region and country, and it is effective for revitalizing the community of a local region [8,9]. To perfect dance techniques, it is essential for a performer to repeatedly practice choreography composed specifically for a song. In general, the performer practices by getting advice on choreography from an instructor and watching sample videos of the choreography [10,11]. However, it is challenging to practice face-to-face because of the spread of COVID-19 infection in recent years. In addition, both the number of instructors and the time available for practice are limited. However, information and communication technology (ICT) and artificial intelligence (AI) technologies can now be used to develop techniques to improve the efficiency of dance practice.

The Yosakoi Soran Festival is a typical Japanese dance event held in Hokkaido. This annual event has an economic effect of more than 20 billion yen (around $173 million USD) because many tourists visit it (https://app.yosakoi-soran.jp/news/view/324, accessed on 16 March 2022). Since Yosakoi Soran [12] is a group dance in which ordinary citizens participate, it is important not only to choreograph individual dances but also to synchronize the troupe members' movements. In particular, abrupt stops during a dance are important in choreography and are especially important for members of a troupe to synchronize with each other. Abrupt stops are an important component not only of Yosakoi Soran but also in traditional Asian dances in China, India, and elsewhere. Figure 1 shows how a stop is performed. Although synchronization of stops among troupe members is key to the successful performance of Yosakoi Soran, it is difficult for performers to notice slight differences in timing between themselves and the other performers. Therefore, it is necessary to construct a dance practice system that detects the timing of stops in a remote environment or in an on-demand format. Our proposal supports individual stopping practice and thus can contribute to improving dance skills.

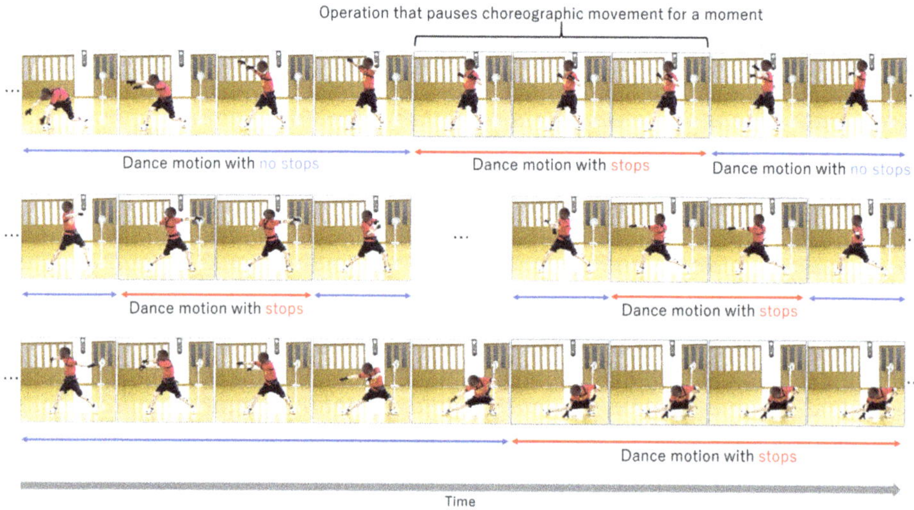

Figure 1. Demonstration of flow of motion in Yosakoi Soran dance. From left to right in the first line: no stop, stops, no stop. The stops factor is incorporated into the continuous dance process as a mid-period.

Various researchers have recently proposed analyses of dance movements using motion capture (MoCap) for teaching and training [13–20]. Those analyses have focused on the performer's pose during a dance. In [13,14], a dance practice support system using MoCap was developed. MoCap-based systems obtain the dance movements of experts and nonexpert performers, calculate the differences between the two, and feed those differences back to the latter. These studies show the effectiveness of using MoCap to support dance practice. In addition, it has been reported that performers prefer dance practices based on an interactive system rather than conventional practice methods such as watching choreography videos and direct instruction by instructors [21]. For Yosakoi Soran in particular, an interactive practice system must help performers synchronize their starts and stops with those of the other troupe members. The importance of synchronized stops during a group dance is shown in [22]. Therefore, it is necessary to analyze the velocities of the joints of a performer during a dance. Moreover, since the choreography of a stop involves the sharpness of movement, analysis focusing on acceleration is also important.

This paper proposes a new stop detection method based on a neural network using a motion sensor's velocity and the acceleration of joint position. An overview of the proposed method is shown in Figure 2. First, the motion data of a Yosakoi Soran performer is obtained by using MoCap. Next, motion features based on velocity and acceleration are calculated from the time-series data of the 3D coordinates of the right fist. Then, a stop is detected based on a feedforward neural network (NN) [23] using those features. Finally, the detected stops are visualized with a 3D humanoid model. The proposed method contributes to the automatic evaluation of Yosakoi Soran movements because it detects the choreographed timing of stops. In addition to streamlining individual repetition practice for Yosakoi Soran, this method is expected to be adapted for various other dance forms that use stops.

This paper is organized as follows. In Section 2, the measurement method of the dance movement by MoCap is explained. In Section 3, the proposed method is explained. Specifically, the method for calculating features based on velocity and acceleration from dance motions and the method for visualizing detected stops are described. In Section 4, the experiments are described. Specifically, the effectiveness of the proposed method is shown by comparing the accuracy of the detection and the visualization of stops between the proposed method and methods for comparison. In Section 5, the conclusions of this paper and future work are described.

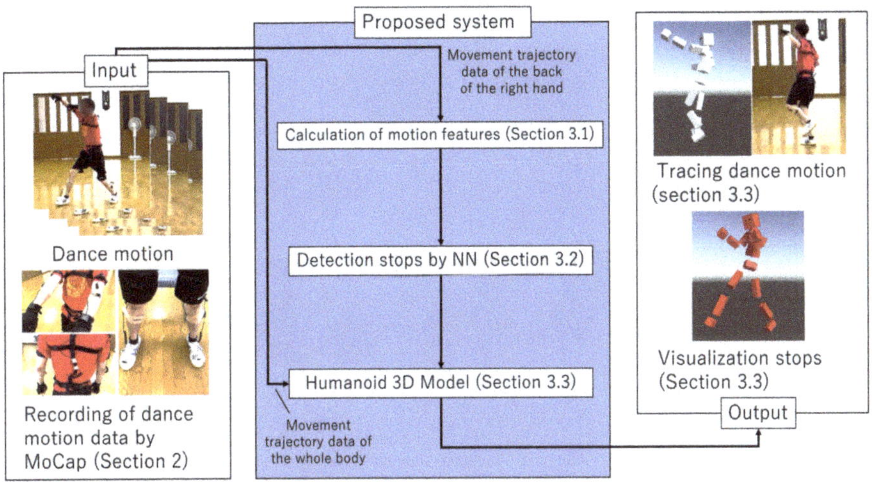

Figure 2. Overview of the proposed system. The dance motion data are recorded by MoCap (Section 2). The proposed system consists of three phases (Section 3). First, the motion features based on velocity and acceleration are calculated (Section 3.1). Stops are detected by a neural network model (Section 3.2). Stops are visualized using a humanoid 3D model via virtual reality spaces (Section 3.3).

2. Recording of Dance Motion Data by MoCap

In this section, we explain how to record a performer's motion data by MoCap to detect stops in Yosakoi Soran. Using Perception Neuron 2.0 (https://www.noitom.com/, accessed on 16 March 2022) (PN), which is NOITOM's motion capture shown in Figure 3, the performer's dance motion is recorded while the performer wears the MoCap. The MoCap consists of multiple small sensors that measure inertia, such as a gyroscope and an accelerometer. Based on the data acquired from these sensors, the posture and position of the performer are estimated.

There are two main advantages of PN. While PN is less accurate than optical MoCaps in estimating posture and position, PN costs only about 1/100 to 1/1000 the cost of an optical MoCap. From this point of view, the practical application of the proposed system becomes realistic. The other advantage of the proposed system is that it has fewer restrictions on its use. For example, an optical MoCap is highly accurate but requires a dedicated studio with multiple cameras and can measure motion only within a limited area. In contrast, PN can be used anywhere if nearby metal or magnetic objects are kept away. On the other hand, PN has some disadvantages: over a long period, errors in a PN gradually increase due to problems in posture- and position-estimation methods. Moreover, the accuracy of position estimation is low. However, if the PN is calibrated periodically and used in a large space where the influence of electromagnetic waves is minimized, the error can be minimized to the extent possible. Actually, Yosakoi musical pieces are short (about 3–10 min each), so the effect of errors caused by continuous use is considered small. Instead, PN is used to analyze various movements such as surgical simulation [24] and the analysis of sports movements [25]. Within this background, we adapt PN for our analysis of dance movement.

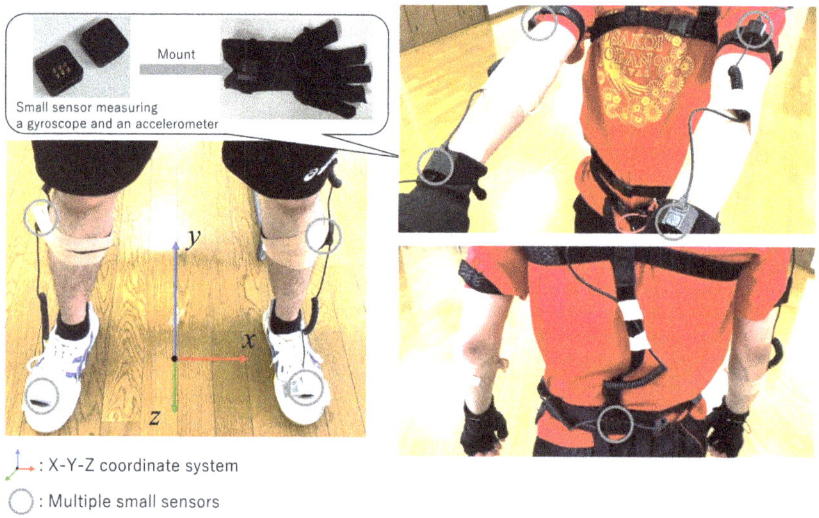

Figure 3. Attachment of PN to the performer is presented with 18 small sensors at hand, arm, shoulder, leg, head, and waist that measure inertia, such as a gyroscope and an accelerometer. The relative positions among the sensors are measured, and the 3D positions of the sensors are obtained.

In dance, the part of the body that has the widest range of movement is the hand. Therefore, in this study, as shown in Figure 3, the MoCap is attached to the performer's hand, and the dance movement, including stops, is recorded. This study describes a method of detecting stops using the movement trajectory data on 3D coordinates of the back of the right hand among the recorded data.

3. Detection and Visualization of Stops by NN Based on Velocity and Acceleration

This section describes the proposed method of detecting stops based on NN and the visualization of detected stops. First, the motion features consist of velocities, and acceleration is calculated using the movement trajectory data of the back of the right hand as recorded by the MoCap. Section 3.1 describes the method of calculating the motion features of a stop. Then, the detection model is constructed based on NN using those features in Section 3.2, and Section 3.3 describes the method of visualizing the detected stops. The details are shown below.

3.1. Calculation of Motion Features

Since stopping is an operation that pauses choreographic movement for a moment, the characteristics of the velocity and acceleration of the hand are calculated. First, we define the movement trajectory data $f_{i,j}(i = 1, 2, \ldots, N; j \in \{x, y, z\} | N$ is the number of series of movement loci) of the back of the right hand from MoCap.

3.1.1. Calculation of Velocity

Since the sensor mounted on the MoCap is extremely sensitive, the obtained movement trajectory data contain high-frequency components such as noise. It is desirable to apply smoothing to the movement trajectory data to remove noise in advance because the calculation of velocity and acceleration requires differentiation of the movement trajectory data. Therefore, the moving average processing [26] is applied to $f_{i,j}$, as shown in Equation (1).

$$f^{\mathrm{ma}}_{i,j} = \frac{1}{2n+1} \sum_{k=i-n}^{i+n} f_{k,j}, \tag{1}$$

where, n is an arbitrary natural number. Note that n and the sampling rate of $f_{k,j}$ were empirically set to 5 (i.e., window size being 12) and 60 Hz, respectively, based on prior equipment preparation for data acquisition. Then, the velocity $f'_{i,j}$ is calculated by the 5-point approximation formula of the derivative in Equation (2).

$$f'_{i,j} = \frac{f^{\mathrm{ma}}_{i-2h,j} - 8f^{\mathrm{ma}}_{i-h,j} + 8f^{\mathrm{ma}}_{i+h,j} - f^{\mathrm{ma}}_{i+2h,j}}{12h}, \tag{2}$$

where, although h is a minute width of differentiation, $h = 1$ is set to obtain the displacement for each frame in this paper. If a high-frequency component is also generated in the calculated velocity $f'_{i,j}$, the velocity $f'_{i,j}$ is applied to the subsequent calculation of the acceleration. Therefore, it is necessary to smooth the transition of $f'_{i,j}$. However, the velocity $f'_{i,j}$ may change drastically, and if moving average processing is applied, the size characteristics may be impaired. Therefore, the following root mean square $f^{\mathrm{rms}}_{i,j}$ is calculated from the velocity $f'_{i,j}$ in Equation (3).

$$f^{\mathrm{rms}}_{i,j} = \sqrt{\frac{1}{2n+1}\sum_{k=i-n}^{i+n}(f'_{k,j})^2}. \tag{3}$$

Then, the one-dimensional velocity f^{abs}_i is calculated from the three-dimensional velocity as shown in Equation (4).

$$f^{\mathrm{abs}}_i = \sqrt{(f^{\mathrm{rms}}_{i,x})^2 + (f^{\mathrm{rms}}_{i,y})^2 + (f^{\mathrm{rms}}_{i,z})^2}. \tag{4}$$

Finally, the change in velocity f^{abs}_i is smoothed by reapplying Equation (2) to the result obtained in Equation (4). Note that the velocity v^{norm}_i is calculated by normalization in the range of 0 to 1. Moreover, the minimum and maximum values used in normalization are determined independently for each set of motion data acquired from each subject. An example of the calculated velocity v^{norm}_i is shown in Figure 4. The green, red, and blue circles indicate frames that have been visually confirmed to be stopped. It can be confirmed that some of the frames whose velocities are minimal or approach 0 are stops (frames of stop possibility, i.e., of the possibility of stopped motion). The green, red, and blue circles are defined as short, normal, and long stops, respectively. Moreover, among the stop possibility frames, a frame that shows no stop is defined as no stop. Table 1 shows the characteristics of each stop. However, since the characteristics of short, normal, and long stops differ, they are labeled separately to improve the NN's accuracy of stop detection below.

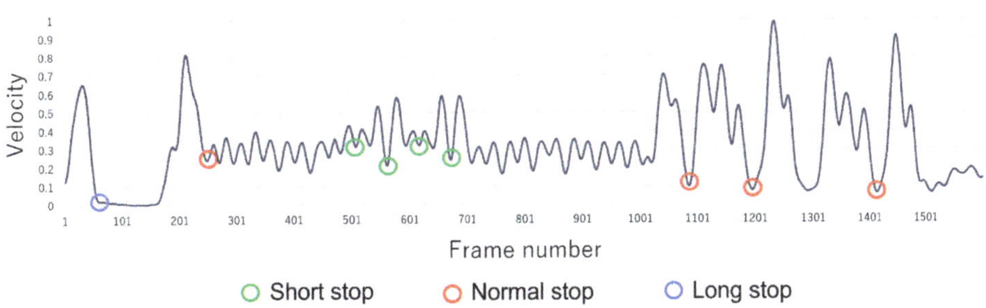

Figure 4. Example of velocity transition.

Table 1. Stops characteristics.

Types of Stops	Details
Short stop	Operation stop time is about 25 frames.
Normal stop	Operation stop time is about 50 frames.
Long stop	Operation is stopped for a while.

3.1.2. Calculation of Acceleration

To detect stopping, we calculate acceleration, which is strongly related to the sharpness of movements. Specifically, using the calculated velocity v_i^{norm}, the first derivative a_i of the velocity v_i^{norm} is calculated by the 5-point approximation formula, as shown in Equation (5).

$$a_i = \frac{v_{i-2h}^{\mathrm{norm}} - 8v_{i-h}^{\mathrm{norm}} + 8v_{i+h}^{\mathrm{norm}} - v_{i+2h}^{\mathrm{norm}}}{12h}, \tag{5}$$

where, to obtain the displacement for each frame, $h = 1$ is set in this paper. Then, a_i^{norm} is calculated by normalizing the result obtained by Equation (5) in the range of -1 to 1, and the minimum and maximum values are determined the same way as the calculation of v_i^{norm}. At this time, the frame at the moment when a_i^{norm} becomes a value from negative to zero or more is the frame when velocity v_i^{norm} becomes the minimum. However, the frame at the moment when the velocity v_i^{norm} approaches zero is slightly delayed with this method. Therefore, the acceleration a_i^{fil} is calculated from a_i^{norm} as shown in Equation (6).

$$a_i^{\mathrm{fil}} = \begin{cases} 0 & \text{if } -0.05 \leq a_i^{\mathrm{norm}} \leq 0.05, \\ a_i^{\mathrm{norm}} & \text{else.} \end{cases} \tag{6}$$

Note that the threshold (=0.05) is set to suppress minute discrepancies, determined empirically from test data from multiple experiments. By calculating a_i^{fil}, it is possible to find all stop possibility frames. An example of the acceleration a_i^{fil} is shown in Figure 5. The green, red, and blue circles are frames whose accelerations change from negative to 0 and which correspond to each stop or stop possibility frame in Figure 4.

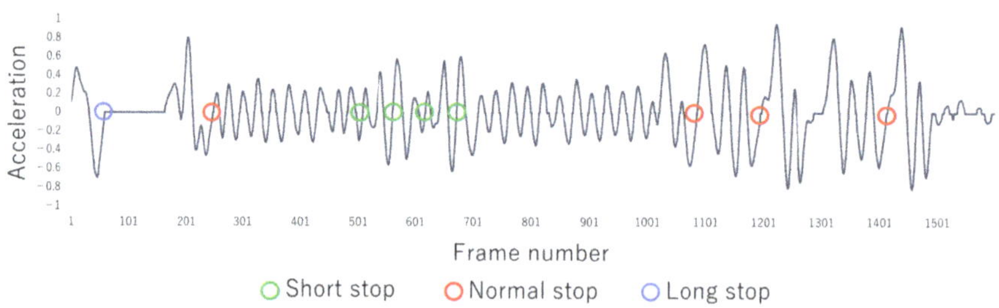

Figure 5. Example of acceleration transition.

3.1.3. Construction of Training Data Set for Stop Detection

The NN's training data are constructed by using velocity and acceleration. The frame number of each stop possibility frame in Figures 4 and 5 is called t. In addition, the operation's stop time tended to be about 25 to 50 frames. Therefore, the characteristics related to before and after the stop possibility frame t are calculated from Equations (7)–(10) by using v_i^{norm} and a_i^{fil}. As a result, the NN's training data for four elements $\boldsymbol{r}_t = [V_{\mathrm{back}}(t), V_{\mathrm{forward}}(t), A_{\mathrm{back}}(t), A_{\mathrm{forward}}(t)]^{\mathrm{T}}$ are obtained. These are approximations of the graph area of Figures 4 and 5 in the range of 25 frames before and after the t frame.

$$V_{\text{back}}(t) = \frac{1}{2}\sum_{k=t-25}^{t-1}(v_k^{\text{norm}} + v_{k+1}^{\text{norm}}), \tag{7}$$

$$V_{\text{forward}}(t) = \frac{1}{2}\sum_{k=t}^{t+24}(v_k^{\text{norm}} + v_{k+1}^{\text{norm}}), \tag{8}$$

$$A_{\text{back}}(t) = \frac{1}{2}\sum_{k=t-25}^{t}(a_k^{\text{fil}} + a_{k+1}^{\text{fil}}), \tag{9}$$

$$A_{\text{forward}}(t) = \frac{1}{2}\sum_{k=t}^{t+25}(a_k^{\text{fil}} + a_{k+1}^{\text{fil}}). \tag{10}$$

3.2. Construction of a Stop Detection Model of the NN

This section describes how to detect a stop by forwarding the propagation type of the NN using the motion features calculated in the previous section. The NN used in the proposed method consists of an input layer, a hidden N-layer, and an output layer. The motion features of each frame calculated in the previous section are used as a vector with $\boldsymbol{r}_t$ as input data. In the output, the values $c = 1, \cdots, C$ of the class indicating the detection result are associated. Let the number of nodes in the input, hidden, and output layers be $M, L,$ and C, respectively. The bias is set by $x_0 = 1$. Let the vector of motion features $\boldsymbol{r}_t$ correspond to $(x_1, \cdots, x_M)$. The following propagation equation is obtained at the node of the l in the middle layer, as shown in Equation (11).

$$h_l(\boldsymbol{r}_t) = h\left(\sum_{m=0}^{M} w_{l,m}^1 x_m\right), \tag{11}$$

where, $w_{m,l}^1$ is the weight between the input layer and the hidden layer. Note that the rectified linear unit (RELU) function [27] is used as the activation function $h(\cdot)$. The linear sum $y_c(\boldsymbol{r}_t)$ of the output of the hidden element, including the bias and the weight $w_{l,c}^2$ between the hidden and output layers, is obtained as shown in Equation (12). $y_c(\boldsymbol{r}_t)$ is the c-th output value in the output layer.

$$y_c(\boldsymbol{r}_t) = \sum_{l=0}^{L} w_{c,l}^2 h_l(\boldsymbol{r}_t). \tag{12}$$

Moreover, the following s_c can be obtained as the probability value indicating the class by conversion with Equation (13) (softmax function [28]) using all the outputs in the output layer.

$$s_c = \frac{\exp\{y_c(\boldsymbol{r}_t)\}}{\sum_{c'=1}^{C}\exp\{y_{c'}(\boldsymbol{r}_t)\}}. \tag{13}$$

In the NN's training, the weights and biases for each layer that minimize the cross-entropy error [29] are determined by using the training data set created in the previous section. Finally, the short, normal, long, and no stops are classified by inputting the same test data as the training data format to the trained NN.

3.3. Visualization of a Stop by Human-like 3D Model in Virtual Reality

This section describes the visualization of a stop by the proposed system. The system described below was developed using Unity (https://unity.com, accessed on 16 March 2022), a virtual reality (VR) development environment. Unity is used to visualize motion analysis research [20,30] and is also useful for actual application development. So, we used the development engine of this virtual environment. For effective practice, it is important for the system to visualize the differences in stop timing between the expert and the performer. Our proposed system can easily confirm the detection results, side by side, of stops made by the expert and by the performer. In the following sections, we describe the functions of the visualization system in detail.

First, using the stop detection model constructed in the previous section, we obtain the stop detection results from the motion data for testing. Next, we prepare the two human 3D models shown in Figure 6 in the VR space and adapt the dance motion. By clicking the Load button below the 3D model, the recorded dance motion is applied to each 3D model. By adopting the dance motion to the human 3D model, the model moves in the same way as the dance motion recorded by motion capture and the color of the model changes to red only when a stop is performed during the dance movement, as shown in Figure 7. The details of the various user interfaces (UI) of the visualization system are shown in Table 2. Specifically, the UI allows the user to fast-forward and rewind the motion data at 0.5×, 1×, and 2× speeds. The upper-right corner of the screen displays the current frame number,giving the performer an idea of the timing of the choreography he or she wants to check. One of them is the dance movement of the performer, and the other is the dance movement of the expert. This allows the performer to visually learn the difference in posture and timing of the stops of the performer and the expert.

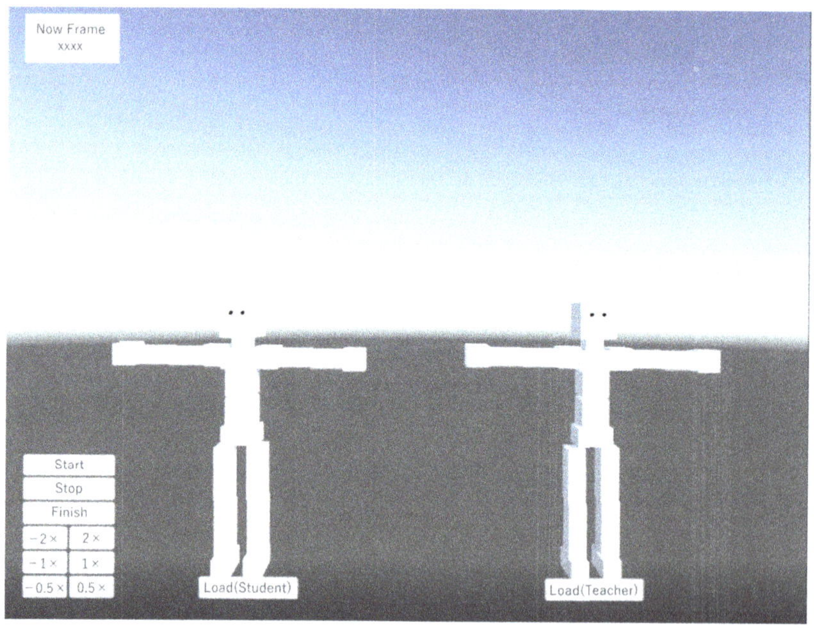

Figure 6. 3D model for stop visualization system.

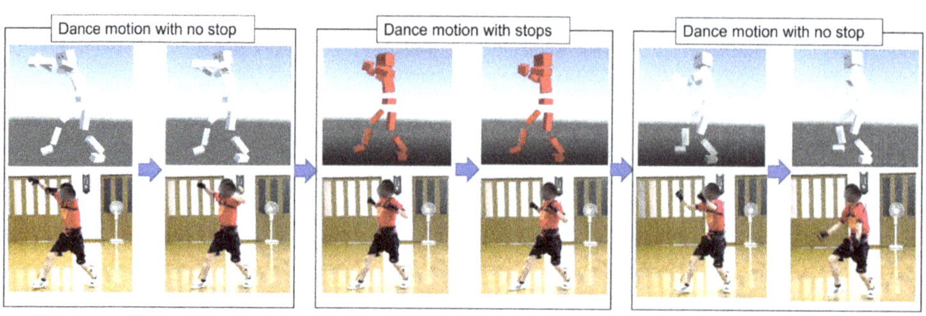

Figure 7. Motion data playback.

Table 2. Details of each UI in visualization system.

UI	Details
Now Frame	Display the current frame.
Load (Student)	Read the dance movement data of the expert dancer.
Load (Teacher)	Read the performer's dance movement data.
Start	Start playback of dance movement data.
Stop	Pause playback of dance movement data.
Finish	End playback of dance movement data.
0.5×	Play back dance movement data at 0.5× speed.
−0.5×	Rewind dance movement data at 0.5× speed.
1×	Play dance movement data at 1× speed
−1×	Rewind dance movement data at 1× speed
2×	Play back dance movement data at 2× speed.
−2×	Rewind dance movement data at 2× speed.

4. Experimental Section

This section verifies the effectiveness of stop detection by the proposed method. It also verifies the visualization accuracy of detected stops.

4.1. Verification of Stop Detection Accuracy

In this section, the accuracy of the proposed method in detecting stops is verified by using comparative methods.

4.1.1. Verification of the Effectiveness of the Proposed Method

In this experiment, the stop detection accuracy is verified. First, the data used in this experiment are described. To construct a high-quality dataset of Yosakoi Soran motion in the first 90 s of the song "Yochore", a total of five dance movements performed by three experienced dancers were recorded and used in the experiment. The details of each dance movement and about the three expert dancers (Subject A–Subject C) are shown in Table 3. In the experiment, we confirmed that stable measurement was achieved with data acquired multiple times by calibrating the PN in a space free of electromagnetic radiation. The choreography of "Yochore" used in the experiment is available on YouTube (https://www.youtube.com/watch?v=FRMOpCPw2xA&t=0s, accessed on 16 March 2022). The number of expert performers was set to an appropriate value based on the [13–15,17,20,31], which performed the motion analysis. The choreography of the Yosakoi dance was composed for ordinary people to earn and reproduce the characteristics of each Yosakoi team, and many various dance processes have been generated at the festival. It is not easy to correct many samples for a partial process of each dance. According to the team's situation of the Yosakoi festival, 679 data were prepared for the detection of the stops in the dance process.

Table 3. Details of each dance movement.

Subject	Gender	Age	Height	Dance Experience	No. of Samples
A	Male	23 years old	164 cm	9 years	128
B1	Male	23 years old	168 cm	15 years	135
B2	Male	23 years old	168 cm	15 years	138
B3	Male	23 years old	168 cm	15 years	135
C	Female	36 years old	164 cm	19 years	143

To evaluate the stop detection accuracy, six methods (Comp. 1–Comp. 6) were used to compare the stop detection accuracy with that of the proposed model (PM). The outline of each method is shown in Table 4. Note that the batch size in the proposed method is set to 581 and the training rate is set to 0.001. The parameters of each comparative method were determined experimentally. Since long short-term memory (LSTM) [32] is analyzing series data in generally, the input data shapes in Comp. 2 and Comp. 3 are changed. Specifically, backwards and forwards are combined in the time direction for the velocity and acceleration features calculated from Equations (7)–(10), respectively. Moreover, the same preprocessing is performed on the time-series data. For this reason, the temporal window size of LSTM is 2, and the number of input nodes is half that of the NN. The effectiveness of feature data is examined by using time-series data. Specifically, the effectiveness of the NN at detecting stops is shown by using LSTM, which is said to be effective for time-series analysis. The effectiveness of the PM is shown below.

- The effectiveness of the feature data is confirmed by comparing the detection accuracy of the stops by the PM and Comp. 1.
- The effectiveness of stop detection by the NN is confirmed by comparing the detection accuracy of stops by the PM and Comp. 2.
- The effectiveness of combining feature data and NN in detecting stops is confirmed by comparing the accuracy of the PM and Comp. 3.
- We verify that the PM is more effective than traditional supervised learning-based methods by comparing the detection accuracy of stops by the PM and Comp. 4–Comp. 6.

Based on the previous section, 679 feature data are constructed from these dance motion data and used as training data for the PM, Comp. 2, and Comp. 4–Comp. 6. Moreover, in the velocity v_i^{norm} and acceleration a_i^{fil} in the previous section, let t be the stop possibility frame in Figures 4 and 5. In this case, 679 time-series data consisting of v_i^{norm} and a_i^{fil} corresponding to the interval of $t-25 \leq i \leq t+25$ are used as training data for Comp. 1 and Comp. 3. Moreover, the time-series data are divided into training data, verification data, and test data as shown in Table 5.

4.1.2. Explanation of Evaluation Index

Next, the evaluation index is explained. In stop detection, it is important to detect a greater number of correct stops. For this reason, it is necessary to evaluate the recall for stop detection. On the other hand, it is also important to reduce false positives. Therefore, it is necessary to evaluate precision as well. For this reason, the $\mathrm{F-measure}$, which is the harmonic mean of precision and recall, is used as the evaluation index for stop detection. The goal of the experiments is to detect stops, and it is impossible to classify short, normal, and long stops accurately, but this does not matter. Therefore, short, normal, and long stops are collectively referred to as stops. At this time, the $\mathrm{F-measure}$ is calculated by Equations (14)–(16).

$$\mathrm{Precision} = \frac{\mathrm{TP}}{\mathrm{TP}+\mathrm{FP}} \tag{14}$$

$$\mathrm{Recall} = \frac{\mathrm{TP}}{\mathrm{TP}+\mathrm{FN}} \tag{15}$$

$$\mathrm{F-measure} = \frac{2 \times \mathrm{Precision} \times \mathrm{Recall}}{\mathrm{Precision}+\mathrm{Recall}} \tag{16}$$

where, true positive (TP) is a set composed of the predictions (stops) and correct answers (stops); false positive (FP) is a set composed of the prediction (stops) and correct answer (no stop); and false negative (FN) is a set composed of the prediction (no stops) and correct answer (stops). The four stop classes are prepared in our detection method via NN architecture. According to our basic concept of detecting the stops in Yosakoi dance, the evaluation is based on two classes: no stop and stops.

Table 4. Overview of each method.

	Model	Data	Input Size	No. of Hidden Layers	No. of Hidden Nodes	Output Size
PM	NN	Feature data	4	1	16	4
Comp. 1	NN	Time-series data	102	1	128	4
Comp. 2	LSTM [32]	Feature data	2 × 2	1	16	4
Comp. 3	LSTM [32]	Time-series data	51 × 2	1	128	4
	Model	**Data**	**Input Size**	**Kernel**		**Output Size**
Comp. 4	Nonlinear Support Vector Machine [33]	Feature data	4	Radial basis function		4
	Model	**Data**	**Input Size**	**No. of Neighbors**		**Output Size**
Comp. 5	k-Nearest Neighbor [34]	Feature data	4	5		1
	Model	**Data**	**Input Size**	**No. of Trees in the Forest**		**Output Size**
Comp. 6	Random Forest [35]	Feature data	4	115		4

Table 5. Breakdown of the number of data sets for each training.

	Training	Verification	Testing
short stop	50	10	10
normal stop	51	12	12
long stop	9	3	3
no stop	363	78	78
Total	473	103	103

4.1.3. Results and Discussion

The stop detection results of all methods are shown in Table 6. The accuracy is based on the value of the test data when the loss of the validation data stops decreasing. The transition of the loss function of the training data and the verification data in the NN-based methods is shown in Figure 8. Note that all data were recognized as no stop, so the precision and F-measure are not calculated in Comp. 3.

Table 6. Stop detection accuracy of each method.

	Precision	Recall	F-Measure
PM	0.938	**0.600**	**0.732**
Comp. 1	**1.000**	0.400	0.571
Comp. 2	0.813	0.520	0.634
Comp. 3	-	0	-
Comp. 4	0.698	0.539	0.546
Comp. 5	0.715	0.472	0.536
Comp. 6	0.641	0.575	0.558

First, from Table 6, the F-measure of the PM stops is shown to be higher than those in Comp. 1–Comp. 6. From this, it can be said that the PM is effective in stop detection. Specifically, first, it can be seen that the PM has succeeded in improving the recall and F-measure when compared with Comp. 1. This shows the effectiveness of introducing feature data. Next, by comparing the PM with Comp. 2, it can be seen that the recall, precision, and F-measure have been successfully improved. This indicates the effectiveness of introducing the NN in stop detection. Next, by comparing the PM with Comp. 3, it can be seen that the recall, precision, and F-measure have been successfully improved. This indicates the effectiveness of combining feature data and the NN in stop detection. Moreover, by comparing the PM and Comp. 2 with Comp. 4–Comp. 6, it can be seen that the precision and F-measure have been successfully improved. This indicates the effectiveness of the NN-based method in stop detection. From the above, the effectiveness of the PM was shown. From Figure 8, it can be seen that the learning stopped before the tendency of the loss function of the verification data began to increase. From this, it can be said that overfitting was successfully suppressed. From Table 6, when feature data are used in the same machine learning method, the precision, recall, and F-measure tend to be higher than when time-series data are used. From this, it can be said that the feature data can express the difference between stops and no stops better than the time-series data. From the PM results, it can be concluded that the most accurate stop detection model has been constructed from among the tested methods.

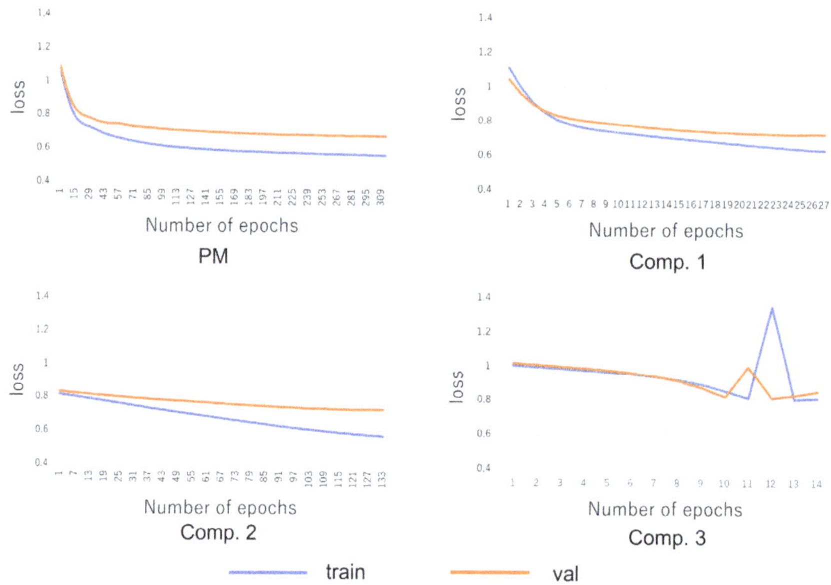

Figure 8. Transition of loss function.

4.2. Confirmation of Visualization Timing of Stops

In this section, the visualization timing of the detected stops is confirmed.

4.2.1. Confirmation Method

The method by which we confirm the visualization accuracy of stops is explained. By applying the various methods used in the previous section to actual dance motion data, the visualization accuracy of the stops is evaluated. The dance motion data to be used in this experiment are 90 s of dance motion performed by another skilled dancer with 15 years of dance experience. This dance motion includes 32 stops (14 short, 15 normal, and 3 long stops). These dance motion data show that various models visualize only the stops performed by the right fist.

At this time, the number of times a correct stop is visualized is the number of detections, the number of times a no stop is visualized is the number of false positives, and the number of times a correct stop is not detected is the undetected number. The visualization accuracy of stops is quantitatively evaluated based on the number of detected stops, the number of false positives, and the number of undetected stops.

4.2.2. Results and Discussion

The visualization result of the stops is shown in Figure 9. However, the green part in the figure is a visualized short stop, the red part is a visualized normal stop, the blue part is a visualized long stop, the black part is a visualized no stop, and the gray part is a point in time where neither a stop nor a not stop was visualized. The numbers of stop detections, nondetections, and false positives are shown in Table 7.

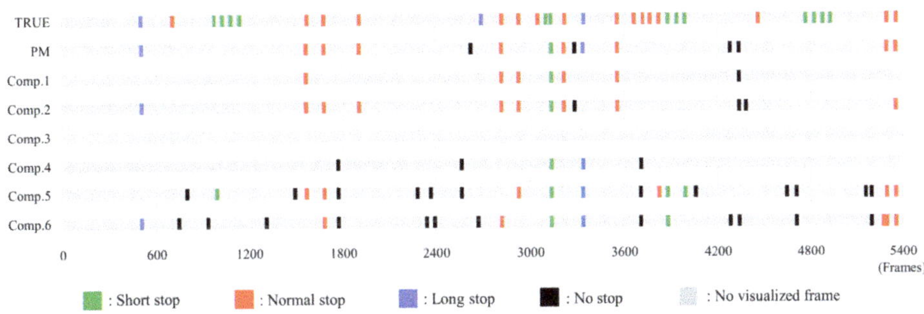

Figure 9. Comparative example of visualization timing of stops.

Table 7. Quantitative evaluation of visualization accuracy.

	No. of Stop Detections	No. of Nondetections	No. of False Positives
PM	**11**	21	4
Comp. 1	6	26	1
Comp. 2	10	22	3
Comp. 3	0	32	0
Comp. 4	2	30	3
Comp. 5	**11**	21	12
Comp. 6	8	24	12

First, Figure 9 and Table 7 show that the PM visualized more stops than Comp. 1–Comp. 4 and Comp. 6. Moreover, the number of false positives in the PM is less than that in Comp. 5. From this, we can judge that the PM is most effective in stop visualization. Then, the characteristics of the stops visualized in the PM are discussed. From Figure 9, a normal stop tends to be more easily visualized than a short or long stop. A normal stop has a longer stopping time than a short stop. These results suggest that the duration of a stop is a major factor in the criteria for judging a stop. In addition, the reason for the low visualization accuracy of a long stop is considered to be the insufficient number of data compared to the other labels.

Next, the false-positive behavior of the PM is discussed. When we checked the falsely detected motions, we found that folding motions such as waving tended to be falsely detected as stops. For example, consider a choreographed movement where the hands are waving. The apparent motion appears to stop only for a short time during the moment when the direction of the wave reverses. In particular, turning movements tended to be similar to those of short stops. From these points, it is considered that folding motions were mistakenly detected as short stops. However, considering the practical application of the stop detection model, the future challenge is how to increase recall while maintaining high precision. This problem is expected to be solved by increasing the number of high-quality data from people with dance experience.

Finally, we discuss aspects of different dance analysis research approaches. Table 8 shows the analysis and visualization functions in the related studies. As the table shows, all the methods perform movement analysis based on the characteristic elements of each dance. First, the proposed method and those in the literature [15,17,20] perform motion analysis based on features characteristic of a dance. In addition, the proposed method and those in the literature [13,14] provide motion analysis and training applications. From the above, it is important to link motion analysis and visualization technologies to develop from fundamental analysis to practical applications, such as dance practice and stage use.

Table 8. List of analyses and visualization features in related studies.

Literature	No. of Subjects/Dance Genres	Analysis Examples	Application for Motion Visualization
PM	5 / Yosakoi (JPN)	NN-based stop detection	Highlighting a teacher and a student stop with VR
Chan et al. [13]	6 / Hip-hop and a-go-go (USA)	Motion matching from motion database	Highlighting incorrect movement joints with VR
Hachimura et al. [14]	5 / Street dance (USA)	-	Overlay of the computer graphics characteristics of a trainer with AR
Shiratori et al. [15]	2 / Aizu-bandaisan (JPN)	Segmentation of motion sequence based on the music rhythm	-
Yoshimura et al. [17]	5 / Fuji Musume (JPN)	Proposal of coordinate system considering local moving for motion tracking	-
Aristidou et al. [20]	3 / Bachatta dance (DMA)	Proposal of Laban Movement Analysis motion features for Laban	Only playback of tracked motion with VR

5. Conclusions

In group dance, stops are important to improve the synchronization of the troupe's movements. However, due to the effects of COVID-19, school, work, and so on, it is difficult to practice dancing in groups. This study proposes a dance practice support system based on detecting stops in a remote environment. The most accurate stop detection among machine learning methods was achieved by training the NN with velocity and acceleration features. Experiments showed the effectiveness of the proposed method. In addition, the detected stops were visualized using a humanoid 3D model. With these, remote control construction of a future dance practice environment was proposed.

In the experiment, the NN's detected stops more accurately than comparative methods. On the other hand, the comparative methods also showed relatively high detection results. From this point of view, it can be said that it is possible to detect stops using various models, and it is expected that a system can be developed at low cost and have a direct effect on dance practice. In addition, because stops are an essential technique not only in Yosakoi Soran but also in a wide range of various genres of dance, the proposed system can be easily applied to other forms of dance by obtaining training data sets from other experienced dancers.

It is difficult to estimate 3D human pose with high accuracy, including depth, based on video images. Therefore, although MoCap was used in this study, the proposed method can be used if the coordinate information can be obtained in three dimensions, including depth. If video-based 3D human pose estimation technology including depth [36–38] can be put to practical use, a better training environment will be constructed.

Author Contributions: Conceptualization, Y.J., G.S. and H.S.; Data curation, Y.J.; Funding acquisition, G.S. and H.S.; Methodology, Y.J., G.S. and H.S.; Software, Y.J.; Validation, Y.J., G.S. and H.S.; Visualization, Y.J.; Writing—original draft, Y.J.; Writing—review & editing, G.S. and H.S. All authors have read and agreed to the published version of the manuscript.

Funding: This research received no external funding.

Institutional Review Board Statement: All subjects gave their informed consent for inclusion before they participated in the study. The study was conducted in accordance with the Declaration of Helsinki, and the protocol was approved by the ethics committee of Muroran Institute of Technology 16000018.

Informed Consent Statement: Informed consent was obtained from all subjects involved in the study.

Data Availability Statement: Not applicable.

Conflicts of Interest: The authors declare no conflict of interest.

Abbreviations

ICT	Information and Communication Technology
AI	Artificial Intelligence
MoCap	Motion capture
NN	Neural network
PN	Perception Neuron
ReLU	Rectified Linear Unit
VR	Virtual Reality
UI	User Interfaces
PM	Proposed method
LSTM	Long Short-Term Memory
TP	True Positive
FP	False Positive
FN	False Negative

References

1. Lindqvist, G. The relationship between play and dance. *Res. Danc. Educ.* **2001**, *2*, 41–52. [CrossRef]
2. Alpert, P.T. The health benefits of dance. *Home Health Care Manag. Pract.* **2011**, *23*, 155–157. [CrossRef]
3. Ward, S.A. Health and the power of dance. *J. Phys. Educ. Recreat. Danc.* **2008**, *79*, 33–36. [CrossRef]
4. Huddy, A.; Stevens, K. The teaching artist: A model for university dance teacher training. *Res. Danc. Educ.* **2011**, *12*, 157–171. [CrossRef]
5. Pedro, R.; Stevens, K.; Scheu, C. Creating a cultural dance community of practice: Building authentic Latin American dance experiences. *Res. Danc. Educ.* **2018**, *19*, 199–215. [CrossRef]
6. Green, J. Power, service, and reflexivity in a community dance project. *Res. Danc. Educ.* **2000**, *1*, 53–67. [CrossRef]
7. Olvera, A.E. Cultural dance and health: A review of the literature. *Am. J. Health Educ.* **2008**, *39*, 353–359. [CrossRef]
8. Hast, D.E. Performance, transformation, and community: Contra dance in New England. *Danc. Res. J.* **1993**, *25*, 21–32. [CrossRef]
9. Jackson, J.B. The opposite of powwow: Ignoring and incorporating the intertribal war dance in the Oklahoma stomp dance community. *Plains Anthropol.* **2003**, *48*, 237–253. [CrossRef]
10. Van Rossum, J.H. The dance teacher: The ideal case and daily reality. *J. Educ. Gift.* **2004**, *28*, 36–55. [CrossRef]
11. Hong, J.C.; Chen, M.L.; Ye, J.H. Acceptance of YouTube applied to dance learning. *Int. J. Inf. Educ. Technol.* **2020**, *10*, 7–13. [CrossRef]
12. Akiba, F. YOSAKOI SORAN as a site of re-localization and its relationship to Japanese pop culture. In Proceedings of the 13th World Congress of the International Association for Semiotic Studies (IASS/AIS), Kaunas, Lithuania, 26–30 June 2017; pp. 653–661.
13. Chan, J.C.; Leung, H.; Tang, J.K.; Komura, T. A virtual reality dance training system using motion capture technology. *IEEE Trans. Learn. Technol.* **2011**, *4*, 187–195. [CrossRef]
14. Hachimura, K.; Kato, H.; Tamura, H. A prototype dance training support system with motion capture and mixed reality technologies. In Proceedings of the 13th IEEE International Workshop on Robot and Human Interactive Communication, Kurashiki, Japan, 22–24 September 2004; pp. 217–222.
15. Shiratori, T.; Nakazawa, A.; Ikeuchi, K. Detecting dance motion structure using motion capture and musical information. In Proceedings of the 10th International Conference on Virtual Systems and Multimedia, Ogaki, Japan, 17–19 November 2004; Volume 4, pp. 1287–1296.
16. Nakazawa, A.; Nakaoka, S.; Ikeuchi, K.; Yokoi, K. Imitating human dance motions through motion structure analysis. In Proceedings of the IEEE/RSJ International Conference on Intelligent Robots and Systems, Lausanne, Switzerland, 30 September–4 October 2002; Volume 3, pp. 2539–2544.
17. Yoshimura, M.; Murasato, H.; Kai, T.; Kuromiya, A.; Yokoyama, K.; mura, K. Analysis of Japanese dance movements using motion capture system. *Syst. Comput. Jpn.* **2006**, *37*, 71–82. [CrossRef]
18. Solberg, R.T.; Jensenius, A.R. Optical or inertial? Evaluation of two motion capture systems for studies of dancing to electronic dance music. In Proceedings of the Systems, Man, and Cybernetics Conferences, Budapest, Hungary, 9–12 October 2016; pp. 469–474.
19. Camurri, A.; El Raheb, K.; Even-Zohar, O.; Ioannidis, Y.; Markatzi, A.; Matos, J.M.; Morley-Fletcher, E.; Palacio, P.; Romero, M.; Sarti, A.; et al. WhoLoDancE: Towards a methodology for selecting motion capture data across different dance learning practice. In Proceedings of the 3rd International Symposium on Movement and Computing, Thessaloniki, Greece, 5–6 July 2016; pp. 1–2.
20. Aristidou, A.; Stavrakis, E.; Charalambous, P.; Chrysanthou, Y.; Himona, S.L. Folk dance evaluation using laban movement analysis. *ACM J. Comput. Cult. Herit.* **2015**, *8*, 1–19. [CrossRef]
21. Wang, Z. Modern social dance teaching approaches: Studying creative and communicative components. *Think. Ski. Creat.* **2022**, *43*, 100974. [CrossRef]

22. Patrona, F.; Chatzitofis, A.; Zarpalas, D.; Daras, P. Motion analysis: Action detection, recognition and evaluation based on motion capture data. *Pattern Recognit.* **2018**, *76*, 612–622. [CrossRef]
23. Zhang, Z.; Li, J.; Manikopoulos, C.; Jorgenson, J.; Ucles, J. HIDE: A hierarchical network intrusion detection system using statistical preprocessing and neural network classification. In Proceedings of the IEEE Workshop on Information Assurance and Security, St. Petersburg, Russia, 21–23 May 2001; Volume 85, p. 90.
24. Kim, H.S.; Hong, N.; Kim, M.; Yoon, S.G.; Yu, H.W.; Kong, H.J.; Kim, S.J.; Chai, Y.J.; Choi, H.J.; Choi, J.Y.; et al. Application of a perception neuron® system in simulation-based surgical training. *J. Clin. Med.* **2019**, *8*, 124. [CrossRef] [PubMed]
25. Akamatsu, Y.; Maeda, K.; Ogawa, T.; Haseyama, M. Classification of expert-novice level using eye tracking and motion data via conditional multimodal variational autoencoder. In Proceedings of the IEEE International Conference on Acoustics, Speech and Signal Processing (ICASSP), Toronto, ON, Canada, 6–11 June 2021; pp. 1360–1364.
26. Box, G.E.; Pierce, D.A. Distribution of residual autocorrelations in autoregressive-integrated moving average time series models. *J. Am. Stat. Assoc.* **1970**, *65*, 1509–1526. [CrossRef]
27. Agarap, A.F. Deep learning using rectified linear units (ReLU). *arXiv* **2018**, arXiv:1803.08375.
28. Memisevic, R.; Zach, C.; Pollefeys, M.; Hinton, G.E. Gated softmax classification. *Adv. Neural Inf. Process. Syst.* **2010**, *23*, 1603–1611.
29. Kline, D.M.; Berardi, V.L. Revisiting squared-error and cross-entropy functions for training neural network classifiers. *Neural Comput. Appl.* **2005**, *14*, 310–318. [CrossRef]
30. Lin, C.Y.; Yang, Z.H.; Zhou, H.W.; Yang, T.N.; Chen, H.N.; Shih, T.K. Combining leap motion with unity for virtual glove puppets. In Proceedings of the IEEE International Conference on Artificial Intelligence and Virtual Reality (AIVR), Taichung, Taiwan, 10–12 December 2018; pp. 251–255.
31. Guo, H.; Sung, Y. Movement estimation using soft sensors based on Bi-LSTM and two-layer LSTM for human motion capture. *Sensors* **2020**, *20*, 1801. [CrossRef] [PubMed]
32. Siami-Namini, S.; Tavakoli, N.; Namin, A.S. A comparison of ARIMA and LSTM in forecasting time series. In Proceedings of the 17th IEEE International Conference on Machine Learning and Applications (ICMLA), Orlando, FL, USA, 17–20 December 2018; pp. 1394–1401.
33. Cortes, C.; Vapnik, V. Support-vector networks. *Mach. Learn.* **1995**, *20*, 273–297. [CrossRef]
34. Cover, T.; Hart, P. Nearest neighbor pattern classification. *IEEE Trans. Inf. Theory* **1967**, *13*, 21–27. [CrossRef]
35. Breiman, L. Random forests. *Mach. Learn.* **2001**, *45*, 5–32. [CrossRef]
36. Cao, Z.; Hidalgo, G.; Simon, T.; Wei, S.E.; Sheikh, Y. OpenPose: Realtime multi-person 2D pose estimation using part affinity fields. *IEEE Trans. Pattern Anal. Mach. Intell.* **2019**, *43*, 172–186. [CrossRef]
37. Chen, C.H.; Ramanan, D. 3d human pose estimation= 2d pose estimation+ matching. In Proceedings of the IEEE Conference on Computer Vision and Pattern Recognition, Honolulu, HI, USA, 21–26 July 2017; pp. 7035–7043.
38. Wang, L.; Chen, Y.; Guo, Z.; Qian, K.; Lin, M.; Li, H.; Ren, J.S. Generalizing monocular 3d human pose estimation in the wild. In Proceedings of the IEEE/CVF International Conference on Computer Vision (ICCV) Workshops, Seoul, Korea, 27 October–2 November 2019.

sensors

MDPI

Article

Proposal of an Alpine Skiing Kinematic Analysis with the Aid of Miniaturized Monitoring Sensors, a Pilot Study

Caterina Russo *, Elena Puppo, Stefania Roati and Aurelio Somà

Department of Mechanical and Aerospace Engineering, Politecnico di Torino, Corso Duca degli Abruzzi 24, 10129 Torino, Italy; elena.puppo@gabel.it (E.P.); stefania.roati@gabel.it (S.R.); aurelio.soma@polito.it (A.S.)
* Correspondence: caterina.russo@polito.it

Abstract: The recent growth and spread of smart sensor technologies make these connected devices suitable for diagnostic and monitoring in different fields. In particular, these sensors are useful in diagnostics for control of diseases or during rehabilitation. They are also extensively used in the monitoring field, both by non-expert and expert users, to monitor health status and progress during a sports activity. For athletes, these devices could be used to control and enhance their performance. This development has led to the realization of miniaturized sensors that are wearable during different sporting activities without interfering with the movements of the athlete. The use of these sensors, during training or racing, opens new frontiers for the understanding of motions and causes of injuries. This pilot study introduced a motion analysis system to monitor Alpine ski activities during training sessions. Through five inertial measurement units (IMUs), placed on five points of the athletes, it is possible to compute the angle of each joint and evaluate the ski run. Comparing the IMU data, firstly, with a video and then proposing them to an expert coach, it is possible to observe from the data the same mistakes visible in the camera. The aim of this work is to find a tool to support ski coaches during training sessions. Since the evaluation of athletes is now mainly developed with the support of video, we evaluate the use of IMUs to support the evaluation of the coach with more precise data.

Keywords: monitoring system; alpine skiing; micro-electro-mechanical system sensors; sport engineering; training

Citation: Russo, C.; Puppo, E.; Roati, S.; Somà, A. Proposal of an Alpine Skiing Kinematic Analysis with the Aid of Miniaturized Monitoring Sensors, a Pilot Study. *Sensors* **2022**, *22*, 4286. https://doi.org/10.3390/s22114286

Academic Editor: Vesa Linnamo

Received: 26 April 2022
Accepted: 2 June 2022
Published: 4 June 2022

1. Introduction

The widespread use of wearable devices based on MEMS (micro-electro-mechanical system) sensors creates the possibility of sports monitoring even for non-professional users [1]. In the field of wearable technologies, one of the most popular is the inertial measurements unit (IMU). These devices let us measure a significant amount of information such as accelerations and angular velocity of a body applying only one sensor and allow us to compute the absolute and the relative angles of each application point [2,3]. For these reasons IMU are widely employed in different fields for diagnostic and monitoring purposes. With a miniaturized monitoring device, it is possible to reach a deep knowledge of the sports activity studied and to have the capability to understand the causes of injuries and adopt strategies to avoid them [4,5]. The data from the monitoring system provides the opportunity to track errors and enhance performance [6–8]. Error detection could help prevent and avoid injuries, help users in a rehabilitation program and help athletes improve their technique [9,10]. Especially in the field of sport and rehabilitation, these sensors are used to better understand the movements of athletes [11–14]. These sensors are also extensively studied in the literature combined with electromyography or video recording to ensure complete data in order to understand athletes' movements and provide them with useful knowledge to improve their performance [15,16]. The number of articles on monitoring technologies to better understand athletes' movements is constantly growing. Particularly noteworthy are the ones regarding the role of wearable technology in the understanding of the kinematics of a sport activity [17], not only in alpine skiing but also in other sport activities such as Nordic Walking; on that topic, see [18] and on cross-country skiing, see [19].

With regards to alpine skiing, it is very important to monitor and evaluate various parameters such as orientation, speed and also the possibility of recognizing the skier's level [20–24]. IMUs are often used in combination with other sensors to ensure a better understanding of movements but above all to correct drift with sensor fusion algorithms. In alpine skiing, normally, the sensors used in conjunction with the IMU are the pressure or force sensors to monitor foot pressure during a ski run. The pressure on the plantar was used to correlate this data with the motion analysis obtained from the IMU; this information is used to tell an expert skier from a recreational one [25]. Also in the work of Bon et al. [26], pressure sensors in ski boots are used, combined with the Xsens MVN Link inertial suit system. Moreover, turn detection is a fundamental task for evaluating the difference between a carved turn and a drifted one [27,28]. In this research, information on the angular velocity and on the radial force has been added to discriminate between different skill levels of skiers. The GNSS is used as the golden standard for these measures and also for tracking trajectories, and the comparison of GNSS and AdMos (a GNSS receiver from Advanced Sport Instruments) is shown in the work of Jolstad et al. [29]. In the work of Fasel et al. [30], a method was adopted to correct drift in speed measurements by installing a reference magnet of the gate of a ski slope. This method allowed them to obtain the kinematics of the athlete's center of gravity with greater accuracy and precision than a GNSS-based system (global navigation satellite system). This method is suitable for indoor monitoring without the use of GNSS. Another method used to compute the trajectory followed is unmanned aerial vehicle (UAV) and ground cameras supported by the neural network and the correlation filter-based algorithm [31]. In the work of Ruiz Garcia et al. [32], to avoid injuries and improve technique, the acceleration and inclinations of a skier during a turn are evaluated. The IMU angular evaluations are compared with the video reference analysis with optimum correlation results.

Focusing on the research that will be presented, the athletes were monitored with five IMUs. This set of sensors allowed us to evaluate the ski run, the number of curves, the ski boot cuffs, the poles and the lower trunk orientation. The principal aim of this work was to enhance the performance of athletes by supporting trainers with qualitative data.

2. Materials and Methods

2.1. Measurements

The IMUs adopted for this work are five Mbientlab MetaMotionRs [33] that integrate three MEMS sensors: an accelerometer, a gyroscope and a barometer. The IMUs are firmly placed on the skier's body, two of them are placed on the upper part of the rear of the ski boot cuffs, the other two on the poles, under the handle and on the lower trunk. We have chosen these sensor locations because we consider them the minimum number to monitor this complex activity. The ski boot sensors allow us to have the best focus concerning what happens at the level of the skis, the pole sensors allow us to understand the mobility of the upper body and the trunk sensor allows us to monitor the inclinations of the core. More sensors could be useful to obtain more information, for example, on the shoulders or knees, but we decided to install a reasonable number of sensors on the athlete so as not to interfere in his movements. The orientation is kept identical for each sensor to obtain a data set easier to analyze. The x axis points upward, the y axis points to the left and the z axis points back. Figure 1 shows the reference system of the IMUs and the reference of each sensor is collected in the following Table 1:

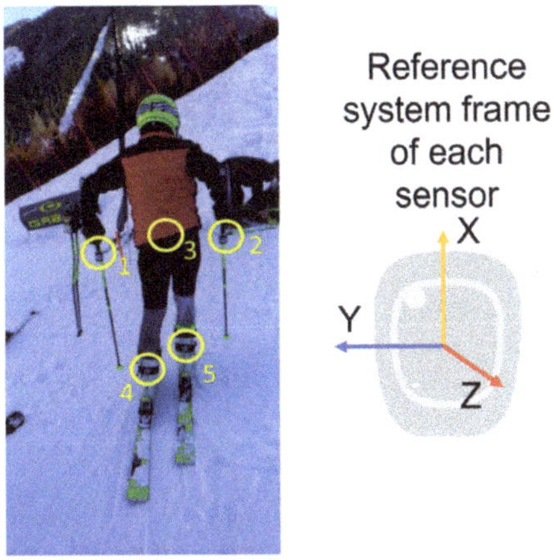

Figure 1. Sensors reference system.

Table 1. Reference number of each sensor.

Number	Name
1	Left pole
2	Right pole
3	Lower trunk
4	Left boot cuff
5	Right boot cuff

The technical specifications of the accelerometer and gyroscope are shown in Tables 2 and 3:

Table 2. Accelerometer technical specification.

Description	Min	Max	Units
Measurement range	±2	±16	g
Resolution	2048	16,384	counts/g

Table 3. Gyroscope technical specification.

Description	Min	Max	Units
Measurement range	±125	±2000	°/s
Resolution	16	262	counts/°

The sampling frequency chosen was 50 Hz because this value allows noise removal and precise results [34]. The value was chosen after different tests both in the laboratory and in the field. We tested different sampling frequencies in order to find the best compromise between the accuracy of the results and the efficiency of the system. As regards the measuring range for the accelerometer, ±16 g and for the gyroscope ±2000°/s were used. For the resolution, the values were, respectively, 2048 counts/g and 16 counts/°. These parameters were identified to obtain a reliable dataset with a sustainable computational time during the post-processing. The data obtained from the IMUs were used to compute the angle of each body segment, relative to its reference frame, on which they were mounted. The barometer measures, sampled at 2 Hz, were used to compute the pressure decreases concerning the slope and these were correlated with the total height of the slope.

2.2. Data Analysis

The experimental activity was developed during ten Giant Slalom and Special Slalom runs by six agonist athletes (one female and five male) and a member of the World Cup team. The mean age of the participants was between 22 and 37. This activity took place in two different locations in Melette (IT) and Pozza di Fassa (IT) during the 2021 winter. The course setting was installed as a distance from the gates of between 18 and 20 m for the Giant Slalom and between 8 and 12 m for the Special Slalom; the lateral distance between the gates was 6–8 m and 2/3 m, respectively. For each sensor, it is possible to evaluate the characteristic angles (roll, pitch and yaw), explained in Table 4, and through these characterize the alpine skiing discipline. In Figure 2, the local reference frame for the boot cuff, the lower trunk and the poles is shown. The x axis is in red, the y axis is in green and the z axis is in blue. In the left-lower corner, the absolute reference frame is shown. The x axis has a leaning angle (16–18°) with respect to the vertical axis, due to the position of the sensors.

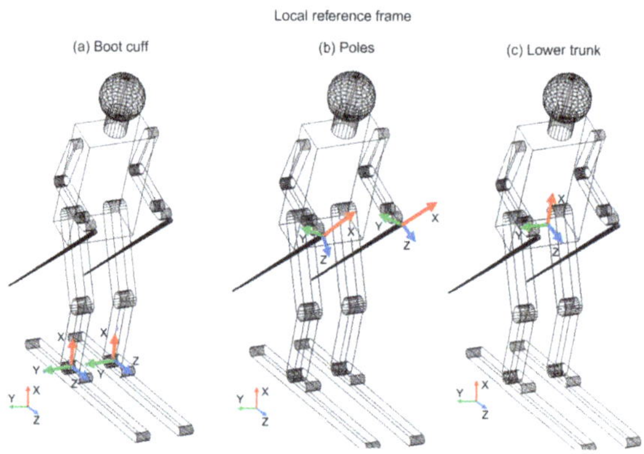

Figure 2. Local reference frame for each monitored part: boot cuff, lower trunk and poles.

Table 4. Description of the characteristic rotations around the sensor axis.

Angles	Description
Roll or Lateral Inclination	Rotation around the z axis
Yaw or turn	Around the x axis
Pitch or flex	Rotation around the y axis

1. The lateral inclination of the boot cuff and the lower trunk, shown in Figure 3, was evaluated with the roll. This represents the typical movement of the skier and can also be considered as an index of the athlete's abilities. The rotation of the pole around the z axis represents the rotation of the pole in the traversal plane, as is visible in Figure 4.

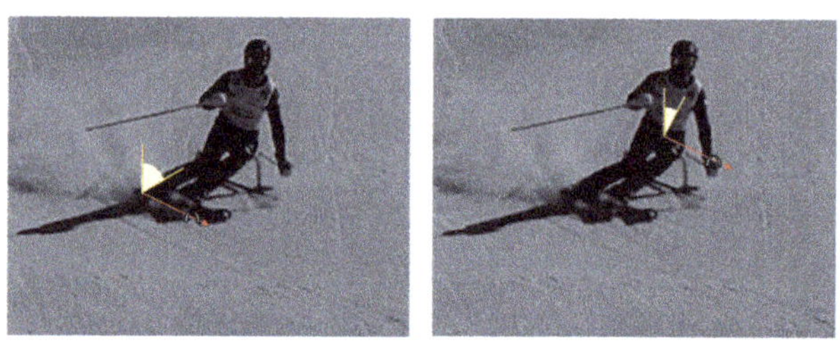

Figure 3. Lateral inclination for boot cuff and lower trunk.

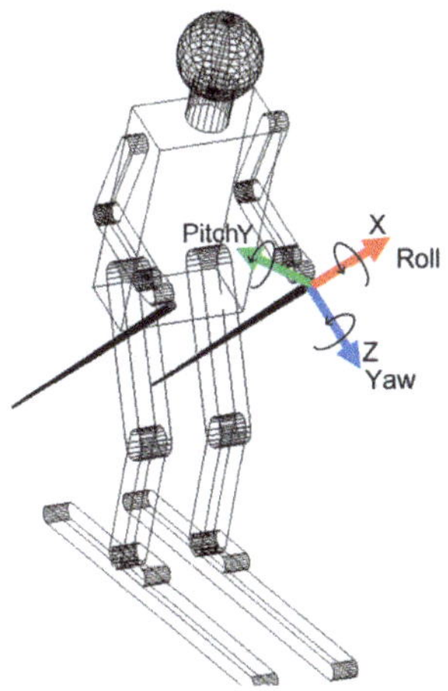

Figure 4. Pole roll, yaw and pitch angles.

2. The yaw angle for boot cuff and lower trunk represents the direction of the ski boot and skies during the ski run and it is directly connected with the trajectory performed. The rotation of the pole around the x axis is expressed by the yaw.
3. With the pitch angle, we measure an angle of a moving part of the ski boot with different stiffness regarding discipline and type of boots. The lower trunk indicates inversion or eversion. In [35], pitch angles varying up to 10° were experimentally observed. This variation changes based on the type of ski boots and on the level of the skier. For agonist skiers, such as the ones tested during the work presented in this paper, the pitch angle variation is reduced. So, in this work, due to the high level of the skier and to the comparison developed, this parameter was neglected. This consideration is not applicable for the poles because their local reference frame is decoupled from the slope and the pitch can be evaluated.

In this section, the algorithm followed to analyzed the data will be presented. The path followed is summarized in the image below (Figure 5):

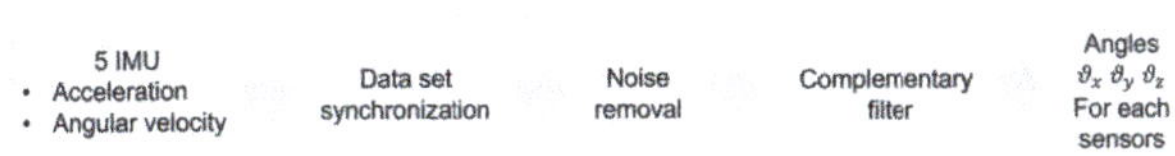

Figure 5. Followed algorithm for the data analysis.

With the complementary filter, the roll and the yaw angles were obtained combining the information of the accelerometer and the gyroscope. Due to the speed during the turns, the large size of the angle θ_{tot} was given by the angular velocity that describes the turns correctly. The information of the accelerometer was used mainly to correct the drifting of the gyroscope. First, the accelerometer signal was filtered with a low-pass filter with a cut-off frequency of 5 Hz; then, the trigonometric formulas were used to compute the orientation values of the body during the ski run. The angle θ_{acc} is evaluated for each axis, measuring the static inclination (Equation (1)). Due to the dynamics of the sport monitored, the angles evaluated with the accelerometer are used only to correct the drift of the angle computed with the gyroscope. In fact, preferably, MEMS (micro-electrical-mechanical system) could be used to measure the static orientation of the sensors, in particular the orientation of an axis with respect to gravity.

$$\theta_{x,acc} = \arctan \frac{a_x}{\sqrt{(a_y^2 + a_z^2)}} \tag{1}$$

$$\theta_{y,acc} = \arctan \frac{a_y}{\sqrt{(a_x^2 + a_z^2)}} \tag{2}$$

$$\theta_{z,acc} = \arctan \frac{a_z}{\sqrt{(a_y^2 + a_x^2)}} \tag{3}$$

The angular velocity, on the other hand, is filtered by a high-pass filter with a cut-off frequency of 0.1 Hz and it is then integrated to calculate the angle.

$$\theta_{gyro} = \int \omega dt \tag{4}$$

So to compute the total angle for each axis, the complementary filter is used:

$$\theta_{tot} = \alpha \cdot \theta_{acc} + \beta \cdot \theta_{gyro} \tag{5}$$

The θ_{tot} computed with this algorithm is the orientation of each body segment monitored during the ski run. The parameters α and β are the coefficients of the filter used to indicate how relevant the angle computed with the gyroscope is with respect to the accelerometer. In this case, due to the high dynamics and velocity of the rotation, the gyroscope measures the orientation of the sensor better, and the orientation computed with the accelerometer is used only to correct the drift. The algorithm used was tested with a calibration phase where the parameters α and β were set to the following values, in accordance with the values in the literature [36,37]: the alpha was equal to 5% and beta to 95%. With these angles, it is possible to obtain different information regarding the run performed, which, combined with the expert assessment of a ski coach, can provide useful corrections for the athlete's technique. This allows the ski coach to obtain a more complete assessment of the skier during the training phase.

The MetaBase application developed by the company Mbientlab [33] was used during the data collection session. Through this application, it is possible to start the data collection of the five sensors at the same time. In order to have a unique starting frame, the accelerations of the poles was used. In particular, in the accelerations along the x axis, the pushing phase during which the athletes pushed themselves off to start the ski run is clearly visible. The three initial peaks in Figure 6 represent the pole pushing.

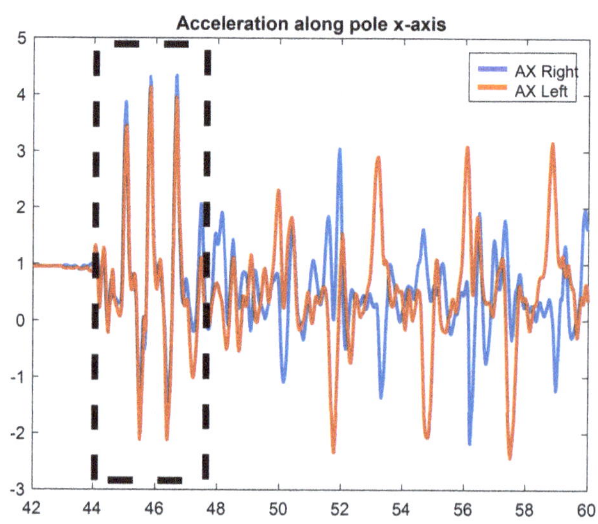

Figure 6. Pole acceleration along x axis.

Then the time of the first peak is taken as the starting point for each run, and all the data collected are cut starting from this reference time.

3. Results

From the post-processed data, it was possible to obtain a large amount of data regarding the ski run, such as the turn definition (number of turns and the average time of each turn) and the average roll and yaw angle for each node monitored (boot cuff, lower trunk and poles).

- **Turn definition**: The beginning and the end of the turn can be defined starting from the edge change visible in the boot roll angle graph. The peaks in this graph represent the maximum lateral inclination of the ski boot during a turn; this occurs in the central phase of the turn when the roll curve change sign corresponds to the edge change and corresponds to the start of the consecutive turn. Hence, each turn can be calculated using zero as a reference, as shown in Figure 7 below.

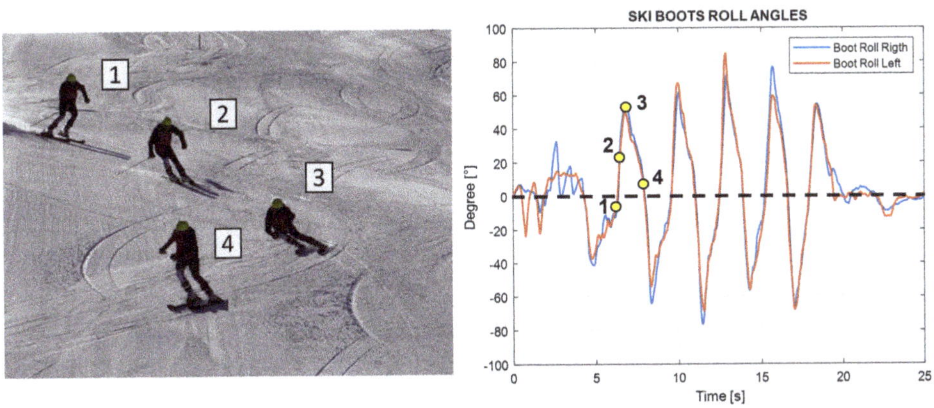

Figure 7. Highlight of one turn in the roll angle graph.

- **Average time of turns**: For each turn, it is possible to compute the time from the start of the turn and its end and compute the average for all of them. This time can be computed considering the starting point when the boot cuff yaw angle is at a maximum, which indicates the edge change as the start of a turn. This yaw angle theoretically should coincide with the zero of the roll angle of the boot cuff. The turns are easily visible in the roll angle graph. The angle value oscillates around the zero mean value; the peaks instead represent the maximum inclination of the skier during the turn. Around the zero value, the ski boot has no inclination and the skier is between two turns; at the peak values, the skier is in the middle of the turn. To know the finish time, the number of peaks in the roll angles plot is counted and compared with the number of gates to validate it.
- **Number of turns**: Counting the peaks from the roll angle graph, the number of turns performed is obtained. The procedure is shown in Figure 8, where on the left the counting peaks and on the right the same turns shown in the ski slope are visible.

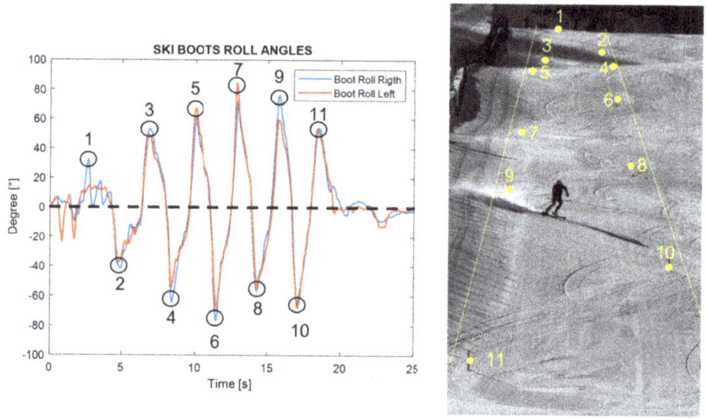

Figure 8. Number of turns in video and roll angle.

- **The average roll angle for ski boots and for the back**: Computing the maximum inclination of the roll angle values, the average lateral inclinations of the skiers during each turn is obtained.

From Figure 9, it can be seen how the mean peak value of the back roll angle always remains lower with respect to the ski boots. This is because the trunk has to remain almost perpendicular to the slope to keep balance and balance the forces during turns. In Table 5, the average values for the roll peak angles for ski boots and back are shown:

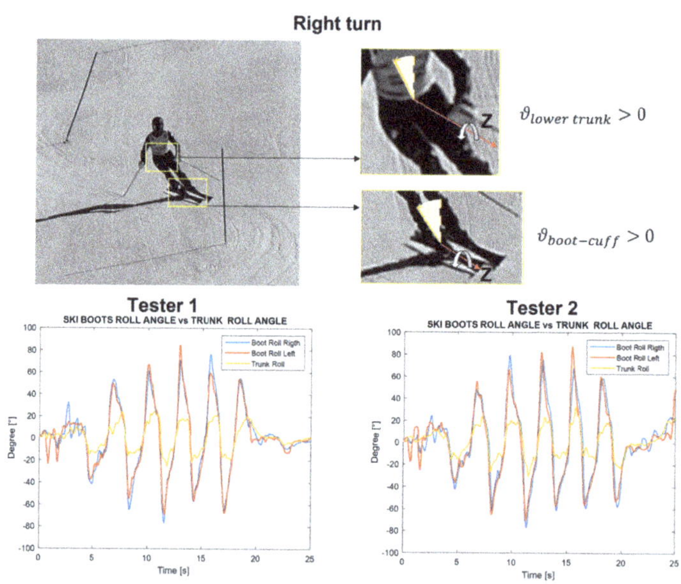

Figure 9. Roll angle for the ski boots and for the back.

Table 5. Average roll angles for ski boots and back.

Tester	Av. Ski Boot Roll Peak Angles	Av. Back Roll Peak Angles
1	62°	23°
2	63°	21°

- **The average yaw peak angle for ski boots and for the back**: The yaw angle values oscillate around zero. The zero position represents the ski orientation alongside the slope; the maximum inclination represents the ski oriented with this angle with respect to the slope. So, in Figure 10, the peaks represent the end of each turn, and the zeros represent the central phase of the turn. In Table 6, the average values of yaw angles are reported.

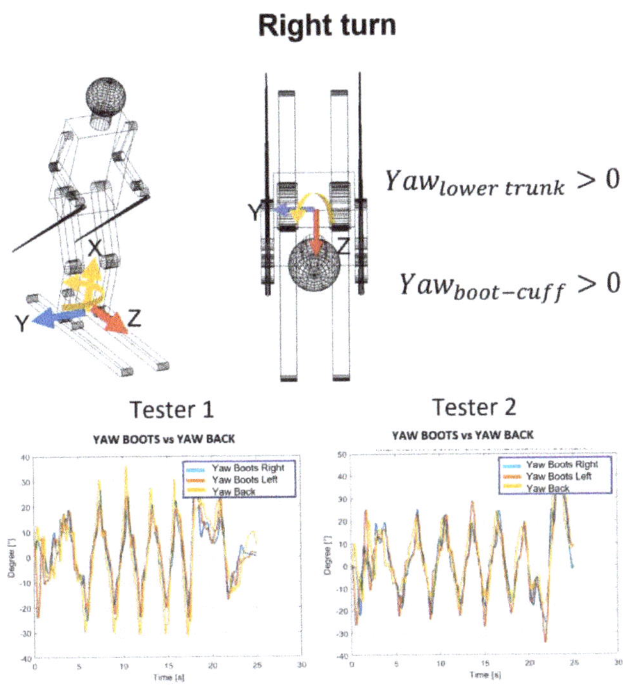

Figure 10. Yaw angle for the ski boots and for the back.

Table 6. Average yaw angles for ski boots and back.

Tester	Av. Ski Boot Yaw Angles	Av. Back Yaw Angles
1	22°	29°
2	21°	29°

- **The average angles for poles**: The poles' orientation became an interesting parameter to understand the correct posture during the ski run. From the angles computed, it is possible to observe the three different movements of the poles. In Figure 11, these movements are shown with the three rotations of the poles. The roll angle of the poles should remain very similar to the lower trunk roll angle because it shows athletes that perform with their arms and poles close to their body, keeping a correct posture. The yaw angle highlights the movement of the poles following the direction of the skies. At the end, the pitch angle emphasizes the understanding of the tendency to approach the gate with arms, which represents imbalanced behavior in the athletes.

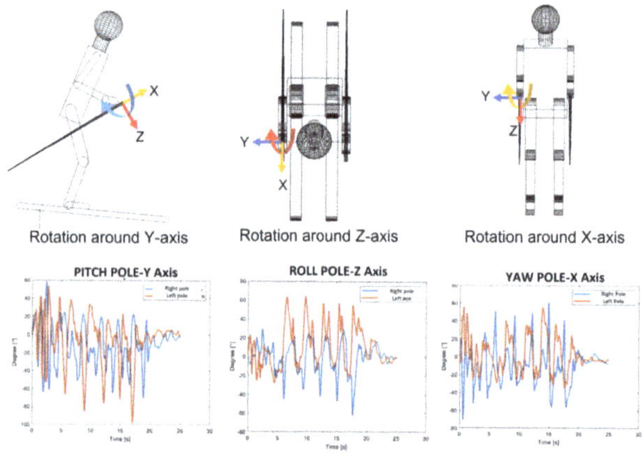

Figure 11. Roll, yaw and pitch angles for poles.

3.1. *Results Interpretation*

In this subsection, the main information extrapolated from these results is explained. The principal aim of a ski coach is to evaluate the ski run, identify the errors and propose a strategy to fix them to the athletes. Due to the high level of ski preparation of the athlete's tested, evident mistakes were not found during this work. At this level, the errors are subtle and should be called inaccuracies. These regard mostly the inclination toward the inner part of the turn and the balancing in the rotation between the trunk and the lower body. For less expert skiers, these inaccuracies became more evident, in particular regarding the position of the body that must lean forward and the central position on the skies. These errors are easily identified by the coach; the actual procedure for the evaluation of the ski run during training is just the comments after a run or the video recording supported by comments from the coach. With the IMU measures, the evaluation could be more complete both for the coach and for the athletes. Therefore, it is mandatory to develop another training session with less-experienced skiers to identify more clearly the differences. Regardless, from the data collected it is possible to identify some behavioral characteristics that are common even among high-level athletes, for example, excessive movements of the upper body, but also some information regarding each turn in order to understand the weaknesses of the ski run and how the athlete has to improve. With the help of these data, the suggestion that the ski coach can offer could be more specific and precise for each athlete. In the case of a less expert skier, for example, if he misses a gate the coach could immediately see the error from the data without seeing the video because there is a missing part in the graph of the roll or yaw of the ski boots. For a more advanced skier, it is possible to evaluate the lateral inclination of each ski boot and the correct use of poles during the ski run. With the data collected on the poles, the possibility of evaluating the amplitude of movements of athletes' arms was observed. The skiing technique requires a very balanced posture, and a strong movement of the upper body and arms can generate an unbalance and consequently a reduction of speed or the incorrect entering in a turn. From the video, it was observed that Tester 1 has evident arm and pole movements during the ski run; in fact, as can be seen in Figure 12, the athlete opens and closes his arms approaching the turn. On the other hand, Tester 2 keeps his arm more still during the entire run. The same can be said from the data collected through the pitch angles; in fact, a more stable value of the peak values of Tester 2 was obtained. In contrast, the graph of the pitch angle for Tester 1 is more segmented and presents more spikes.

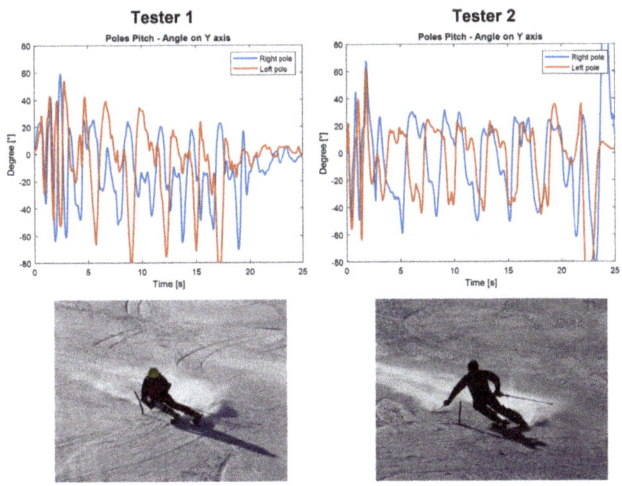

Figure 12. Pitch angle for right and left pole for Testers 1 and 2.

Despite this imbalance in the upper body movements, Tester 1 performs the ski run more efficiently (17.87 s of total time vs 18.29 s for Tester 2), finishing the slope faster. However, in order to improve his technique, it was mandatory that his coach pointed out this behavior in order to fix it. In fact, it would be more useful if the posture that he kept during the turn were maintained during all the slopes to complete a more efficient run. These considerations are fundamental and preparatory to the identification of evaluation indexes. These indexes are useful to evaluate the ski run of each athlete and compare different runs. The speed, the time between two gates and the lateral inclination of the ski boots are usable information in the definition of these indexes. A new test session is mandatory to pursue this purpose. Only in this way is it possible to evaluate whether these indices evaluate the different abilities among athletes.

3.2. *Comparison of Video and IMU Data*

First of all, the IMU data were synchronized using the first pushing phase of each athlete at the starting phase. The pushing phase is clearly visible from the data obtained from the accelerometer placed on the poles, especially along the x axis that is aligned with the longitudinal axis. With the software, Kinovea, the recorded video of the test was analyzed. Each video was synchronized temporally with the data measured with the IMUs using the chronometer tool available in the software and made it start during the first push on the poles. Then, in order to obtain a comparison between the angles measured by the sensors and the ones recorded from the video, a projection of the IMU ones is mandatory. The angles computed with the IMU are in a local reference frame; in contrast, the angles recorded with the camera are in a global reference frame. To compare them, evaluation of the slope is necessary. To obtain the slope, barometric measures were used in combination with the information about the ski setting. In order to compute the slope of each segment of the ski run, we assumed the distance between two gates was about 9 m (ΔZ in Figure 13) and the lateral distance (ΔY) was about 10 m and we measured the height with the barometer (ΔH). This approximation allows us to obtain the rotation matrix for each curve to compare the video data with the IMU ones. The rotation matrices were used to multiply the roll and yaw angle in order to compute the same angles in the global reference frame. The use of the barometer allows a first approximation of these matrices that, in future work, have to be computed with a more accurate method such as GNSS.

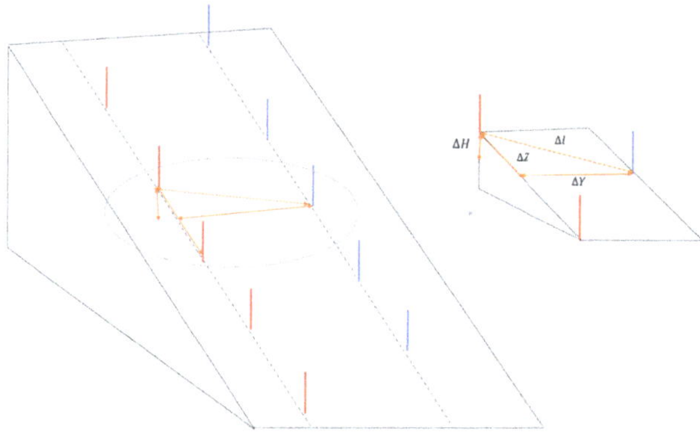

Figure 13. Slope inclination measures.

Once the slopes of each part of the ski run are obtained, it is possible to project the roll and the yaw measured with the IMU in the absolute reference. A good comparison was obtained, giving confidence in the algorithm used. In future work, more accurate validation [38] with 3D motion capture simulation has to be performed in order to be able to also validate the boot cuff angles in the other planes and also the lower trunk and pole angles and to compute the inclination of the slope properly. We show the result obtained for two consecutive turns of a ski run:

Tester 1 was Giant Slalom (see Figure 14).

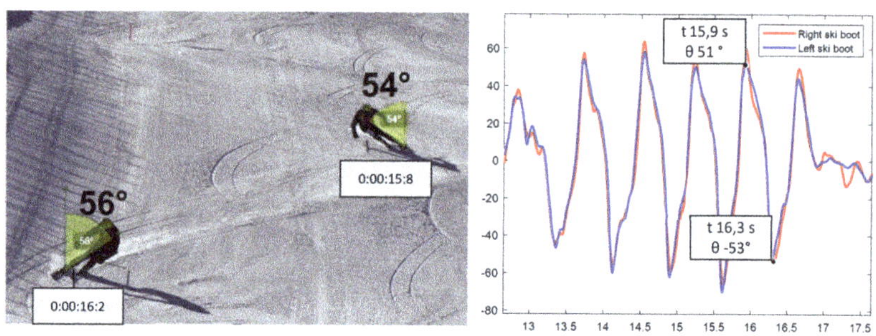

Figure 14. Comparing the video roll angles with the roll angles computed with the IMU.

From the analyzed data, an inclination of 54° at a time of 15.8 s was identified for the left boot during the curve from left to right, which has the left ski as external. An inclination of 56° at the time of 16.2 s was identified in the next turn, with the right ski as external. The results obtained show agreement between the video angles and the IMUs. To also compare the yaw angle and the pole angle, tridimensional information from the video is necessary, which can be obtained only with a deeper video analysis. Performing this comparison enables a better understanding of the values computed with the IMUs that are reasonable in the alpine skiing contest. As mentioned before, a 3D motion-capture video analysis has to be performed in order to obtain more precise data regarding all the three-axis angles, but this comparison offers an index of the right path we are following. In this comparison, the percentage quote of human error, due to the positioning of the sensors and due to the manual insertion of the start timing in the video recording and the inclination of the camera, must be inserted. Moreover, the technical error intrinsic in the measure of an IMU must also be inserted.

4. Discussion

The main aim of this work was proposing a wearable IMU-based monitoring system for the evaluation of the ski run. A monitoring structure, based on different application points, was realized to improve athlete performances, and provide the coach qualitative data to support his evaluation. The monitoring device is meant to be worn during the training sessions and the data collected and the coach's evaluation of these data allow technique improvement. The monitoring devices based on wearable sensors are easily and intuitively worn during a ski run. The number of application points (upper limbs, lower limbs and trunk) gives a complete evaluation of the gesture and the kinematics of each joint. The turn detection in this work is defined using ski boots' roll angles. When the roll angle curve changes sign, this identifies the edge change of the ski and the start of a new turn. In this pilot work, the focus of the work was to develop and find some indication from the sensors that could be used by the trainer to improve his evaluation of an athlete. As Martinez et al. [24,27,28] and Snyder et al. [23] mentioned in their work, the identification of the turn with the roll angle is valid if the parallel turns are performed and if the athlete is an expert. They developed a sophisticated algorithm to identify the turn for different types of turns (parallels, drifted and snowplow) and at the same time to identify the levels of different skiers. Video recordings were used in this work to compare the IMU angles with the video ones but also to start from the coach's standard way of evaluating an athlete. We consider the evaluation of the trajectory not fundamental for the proposed work. The use of GNSS and other methods of localization presented by Jølstad et al. [29] and Fasel et al. [30] is useful if we exclude the video recording. Moreover, the use of pressure sensors, as explained in the work of Bon et al. [26] and Matsumura et al. [25], we consider to provide useful information but not to be fundamental for this pilot work in which we preferred the easy application of five sensors for the kinematics data. With the data acquired, the information that can be extracted from them was shown. Especially with the main angles, which characterize the alpine skiing discipline, it is possible to obtain a variety of information regarding the quality of the ski run. In fact, it is possible to extract some interesting information by analyzing each turn separately and synchronizing the different sensors. This paper showed the roll angle evaluation of the boot cuff and its comparison with the data obtained from the video frames, obtaining a promising agreement concerning these data. The data coming from the poles can also add some interesting information about the arms' and upper limbs' movements during a ski run. This information can be useful to understand the right balance between the upper and the lower part of the body to guarantee an optimal ski run. The limits of this work are mainly due to the reduced number of skiers tested, both in terms of numbers and in terms of ski levels. It would be useful to test a larger group of testers with a high variety of ski levels to be able to distinguish an expert skier from a lower-lever skier. However, a more detailed 3D motion capture validation would be required to complete this work. Introducing, in future work, a comparison between measured data and a simulation of an anthropomorphous multibody model would help in the evaluation of the characteristic angles explained in this work. We also want to consider the possibility of evaluating a skier's experience based on the ski boot roll angles. The higher the roll angles, the higher the expertise level of the skier will be, but we need more data to prove this consideration.

5. Conclusions

In conclusion, the aim of this work is to provide the ski coach and the athlete with a larger amount of data on which to base the evaluation of a ski run. Up until now, the evaluation of a ski run was developed only using video and the information given by a coach; with this monitoring tool, the video information can be correlated with some graphs that can be useful for technique improvement. In particular, it can be useful for observing the unbalance between the upper body and the lower body, for observing whether the athlete reaches the gate with an arm, for seeing if it is necessary to improve the lateral inclination of the skies during a turn and for evaluating the correct timing of a weight movement approaching a turn.

Author Contributions: The authors contributed equally to this manuscript. All authors read and agreed to the published version of the manuscript.

Funding: This research received no external funding.

Institutional Review Board Statement: Not applicable.

Informed Consent Statement: Not applicable.

Data Availability Statement: Not applicable.

Acknowledgments: The research was developed during the cooperation of Politecnico di Torino, Torino-Italy and Gabel Srl, Rosà-Italy. The authors would like to thank Gabel Srl for their important support during the in field test.

Conflicts of Interest: The authors declare no conflict of interest.

References

1. Aroganam, G.; Manivannan, N.; Harrison, D. Review on Wearable Technology Sensors Used in Consumer Sport Applications. *Sensors* **2019**, *19*, 1983. [CrossRef] [PubMed]
2. Chardonnens, J.; Favre, J.; Cuendet, F.; Gremion, G.; Aminian, K. A system to measure the kinematics during the entire ski jump sequence using inertial sensors. *J. Biomech.* **2013**, *46*, 56–62. [CrossRef] [PubMed]
3. Balloch, A.S.; Meghji, M.; Newton, R.U.; Hart, N.H.; Weber, J.A.; Ahmad, I.; Habibi, D. Assessment of a Novel Algorithm to Determine Change-of-Direction Angles While Running Using Inertial Sensors. *J. Strength Cond. Res.* **2020**, *34*, 134–144. [CrossRef] [PubMed]
4. Rawashdeh, S.A.; Rafeldt, D.A.; Uhl, T.L. Wearable IMU for Shoulder Injury Prevention in Overhead Sports. *Sensors* **2016**, *16*, 1847. [CrossRef]
5. Arlotti, J.; Carroll, W.; Afifi, Y.; Talegaonkar, P.; Albuquerque, L.; Burch, R.; Ball, J.; Chander, H.; Petway, A. Benefits of IMU-based Wearables in Sports Medicine: Narrative Review. *Int. J. Kinesiol. Sports Sci.* **2022**, *10*, 36–43. [CrossRef]
6. Tedesco, S.; Alfieri, D.; Perez-Valero, E.; Komaris, D.-S.; Jordan, L.; Belcastro, M.; Barton, J.; Hennessy, L.; O'Flynn, B. A Wearable System for the Estimation of Performance-Related Metrics during Running and Jumping Tasks. *Appl. Sci.* **2021**, *11*, 5258. [CrossRef]
7. Komaris, D.S.; Tarfali, G.; O'Flynn, B.; Tedesco, S. Unsupervised IMU-based evaluation of at-home exercise programmes: A feasibility study. *BMC Sports Sci. Med. Rehabil.* **2022**, *14*, 28. [CrossRef]
8. Mocera, F.; Aquilino, G.; Somà, A. Nordic walking performance analysis with an integrated monitoring system. *Sensors* **2018**, *18*, 1505. [CrossRef]
9. Sands, W.A.; Kavanaugh, A.A.; Murray, S.R.; McNeal, J.R.; Jemni, M. Modern Techniques and Technologies Applied to Training and Performance Monitoring. *Int. J. Sports Physiol. Perform.* **2017**, *12*, 263–272. [CrossRef]
10. Thornton, H.; Delaney, J.; Duthie, G.; Dascombe, B. Developing Athlete Monitoring Systems in Team Sports: Data Analysis and Visualization. *Int. J. Sports Physiol. Perform.* **2019**, *14*, 698–705. [CrossRef]
11. Kiernan, D.; Hawkins, D.A.; Manoukian, M.A.C.; McKallip, M.; Oelsner, L.; Caskey, C.F.; Coolbaugh, C.L. Accelerometer-based prediction of running injury in National Collegiate Athletic Association track athletes. *J. Biomech.* **2018**, *17*, 201–209. [CrossRef] [PubMed]
12. Mooney, R.; Corley, G.; Godfrey, A.; Quinlan, L.; OLaighin, G. Inertial Sensor Technology for Elite Swimming Performance Analysis: A Systematic Review. *Sensors* **2016**, *16*, 18. [CrossRef]
13. Hamidi Rad, M.; Gremeaux, V.; Massé, F.; Dadashi, F.; Aminian, K. SmartSwim, a Novel IMU-Based Coaching Assistance. *Sensors* **2022**, *22*, 3356. [CrossRef] [PubMed]
14. Benson, L.C.; Räisänen, A.M.; Clermont, C.A.; Ferber, R. Is This the Real Life, or Is This Just Laboratory? A Scoping Review of IMU-Based Running Gait Analysis. *Sensors* **2022**, *22*, 1722. [CrossRef]
15. Repnik, E.; Puh, U.; Goljar, N.; Munih, M.; Mihelj, M. Using Inertial Measurement Units and Electromyography to Quantify Movement during Action Research Arm Test Execution. *Sensors* **2018**, *18*, 2767. [CrossRef] [PubMed]
16. Taborri, J.; Keogh, J.; Kos, A.; Santuz, A.; Umek, A.; Urbanczyk, C.; van der Kruk, E.; Rossi, S. Sport Biomechanics Applications Using Inertial, Force, and EMG Sensors: A Literature Overview. *Appl. Bionics Biomech.* **2020**, *2020*, 1176–2322. [CrossRef] [PubMed]
17. Adesida, Y.; Papi, E.; McGregor, A.H. Exploring the Role of Wearable Technology in Sport Kinematics and Kinetics: A Systematic Review. *Sensors* **2019**, *19*, 1597. [CrossRef] [PubMed]
18. Russo, C.; Mocera, F.; Somà, A. MEMS sensors for sport engineer applications. *IOP Conf. Ser. Mater. Sci. Eng.* **2021**, *1038*, 012056. [CrossRef]

19. Bruzzo, J.; Perkins, N.C.; Mikkola, A. Embedded inertial measurement unit reveals pole lean angle for cross-country skiing. *Sports Eng.* **2020**, *23*, 6. [CrossRef]
20. Sulheim, S.; Ekeland, A.; Bahr, R. Self-estimation of ability among skiers and snowboarders in alpine skiing resorts. *Knee Surg. Sports Traumatol. Arthrosc.* **2007**, *15*, 665–670. [CrossRef] [PubMed]
21. Kondo, A.; Doki, H.; Hirose, K. Motion analysis and joint angle measurement of skier gliding on the actual snow field using inertial sensors. *Procedia Eng.* **2013**, *60*, 307–312. [CrossRef]
22. Lee, S.; Kim, K.; Kim, Y.H. Motion Analysis in Lower Extremity Joints During Ski Carving Turns Using Wearable Inertial Sensors and Plantar Pressure Sensors. In Proceedings of the 2017 IEEE International Conference on Systems, Man, and Cybernetics (SMC), Banff, AB, Canada, 5–8 October 2017.
23. Snyder, C.; Martínez, A.; Jahnel, R.; Roe, J.; Stöggl, T. Connected skiing: Motion quality quantification in alpine skiing. *Sensors* **2021**, *21*, 3779. [CrossRef] [PubMed]
24. Martínez, A.; Brunauer, R.; Venek, V.; Snyder, C.; Jahnel, R.; Buchecker, M.; Thorwartl, C.; Stöggl, T. Development and Validation of a Gyroscope-Based Turn Detection Algorithm for Alpine Skiing in the Field. *Front. Sports Act. Living* **2019**, *1*, 18. [CrossRef] [PubMed]
25. Matsumura, S.; Ohta, K.; Yamamoto, S.I.; Koike, Y.; Kimura, T. Comfortable and convenient turning skill assessment for alpine skiers using imu and plantar pressure distribution sensors. *Sensors* **2021**, *21*, 834. [CrossRef] [PubMed]
26. Bon, I.; Očić, M.; Cigrovski, V.; Rupčić, T.; Knjaz, D. What Are Kinematic and Kinetic Differences between Short and Parallel Turn in Alpine Skiing? *Int. J. Environ. Res. Public Health* **2021**, *18*, 3029. [CrossRef]
27. Martínez, A.; Jahnel, R.; Buchecker, M.; Snyder, J.; Brunauer, R.; Stöggl, T. Development of an automatic alpine skiing turn detection algorithm based on a simple sensor setup. *Sensors* **2021**, *19*, 902. [CrossRef] [PubMed]
28. Martínez, A.; Snyder, C.; Moore, S.R.; Stöggl, T. A Comprehensive Comparison and Validation of Published Methods to Detect Turn Switch during Alpine Skiing. *Sensors* **2022**, *21*, 2573. [CrossRef] [PubMed]
29. Jølstad, P.A.; Reid, R.C.; Gjevestad, J.G.; Gilgien, M. Validity of the AdMos, Advanced Sport Instruments, GNSS Sensor for Use in Alpine Skiing. *Remote Sens.* **2022**, *14*, 22. [CrossRef]
30. Fasel, B.; Gilgien, M.; Sporri, J.; Aminian, K. A New Training Assessment Method for Alpine Ski Racing: Estimating Center of Mass Trajectory by Fusing Inertial Sensors with Periodically Available Position Anchor Points. *Front. Physiol.* **2018**, *9*, 1203. [CrossRef]
31. Qi, J.; Li, D.; Zhang, C.; Wang, Y. Alpine Skiing Tracking Method Based on Deep Learning and Correlation Filter. *IEEE Access* **2022**, *10*, 39248–39260. [CrossRef]
32. Ruiz-García, I.; Navarro-Marchal, I.; Ocaña-Wilhelmi, J.; Palma, A.J.; Gómez-López, P.J.; Buchecker, M.; Carvajal, M.A. Development and evaluation of a low-drift inertial sensor-based system for analysis of alpine skiing performance. *Sensors* **2021**, *21*, 2480. [CrossRef] [PubMed]
33. Available online: https://mbientlab.com (accessed on 25 April 2022).
34. Yu, G.; Jang, Y.J.; Kim, J.; Kim, J.H.; Kim, H.Y.; Kim, K.; Panday, S. Potential of IMU Sensors in Performance Analysis of Professional Alpine Skiers. *Sensors* **2016**, *16*, 463. [CrossRef] [PubMed]
35. Petrone, N.; Marcolin, G.; Panizzolo, F.A. The effect of boot stiffness on field and laboratory flexural behavior of alpine ski boots. *Sports Eng.* **2013**, *16*, 265–280. [CrossRef]
36. Gui, P.; Tang, L.; Mukhopadhyay, S. MEMS based IMU for tilting measurement: Comparison of complementary and kalman filter based data fusion. In Proceedings of the IEEE 10th Conference on Industrial Electronics and Applications (ICIEA), Auckland, New Zealand, 15–17 June 2015; pp. 2004–2009.
37. Ariffin, N.H.; Arsad, N.; Bais, B. Low cost MEMS gyroscope and accelerometer implementation without Kalman Filter for angle estimation. In Proceedings of the International Conference on Advances in Electrical, Electronic and Systems Engineering (ICAEES), Putrajaya, Malaysia, 14–16 November 2016; pp. 77–82.
38. Panizzolo, F.A.; Marcolin, G.; Petrone, N. Comparative evaluation of two skiing simulators as functional training devices for recreational skiers. *J. Sports Sci. Med.* **2013**, *12*, 151–158. [PubMed]

sensors MDPI

Article

Physical Demands during the Game and Compensatory Training Session (MD + 1) in Elite Football Players Using Global Positioning System Device

Gabriel Calderón-Pellegrino [1,*], Leonor Gallardo [1], Jorge Garcia-Unanue [1], Jose Luis Felipe [2], Antonio Hernandez-Martin [1], Víctor Paredes-Hernández [3] and Javier Sánchez-Sánchez [2]

1 IGOID Research Group, University of Castilla-La Mancha, 45071 Toledo, Spain; leonor.gallardo@uclm.es (L.G.); jorge.garciaunanue@uclm.es (J.G.-U.); antonio.hmartinsan@uclm.es (A.H.-M.)
2 Faculty of Sport Sciences, Universidad Europea de Madrid, 28670 Madrid, Spain; joseluis.felipe@universidadeuropea.es (J.L.F.); javier.sanchez2@universidadeuropea.es (J.S.-S.)
3 Sport Science Institute, Universidad Camilo José Cela, 28692 Madrid, Spain; vparedes@ucjc.edu
* Correspondence: gcalderonpellegrino@gmail.com; Tel.: +34-952268800; Fax: +34-926295301

Abstract: The aims of this study were to analyze the differences of physical demands of non-starter players regarding the playing time during the competition and to evaluate the physical demands of the compensatory training (MD + 1C) for substitute players in elite football. The match statistics and MD + 1C of substitute players from a professional Spanish LaLiga football club were analyzed using a 10-Hz global positioning system (GPS) Apex GPS system device, which has been validated as a reliable and valid method to analyze performance in team sports, during all games of the 2016/2017, 2017/2018 and 2018/2019 seasons. The starting players showed both lower total distances covered and high-intensity actions compared to the substitutes. Regarding the minutes played by the substitutes, greater physical performance was found for the players with fewer minutes (5–15 min). Furthermore, no differences were found between first and second divisions regarding physical performance of substitutes ($p > 0.05$). This study highlights the importance of individualizing the workload of training sessions for substitutes and starters. Furthermore, the complementary session should be individualized according to the minutes played by the substitutes. These players are potentially under-loaded compared to starters, especially in terms of high-intensity actions, therefore additional session-specific training for each substitute would be useful to reach the optimal training load according to the minutes played during the game.

Keywords: global positioning system; tracking system; quantification; substitutes; high-intensity; compensatory training

Citation: Calderón-Pellegrino, G.; Gallardo, L.; Garcia-Unanue, J.; Felipe, J.L.; Hernandez-Martin, A.; Paredes-Hernández, V.; Sánchez-Sánchez, J. Physical Demands during the Game and Compensatory Training Session (MD + 1) in Elite Football Players Using Global Positioning System Device. *Sensors* **2022**, 22, 3872. https://doi.org/10.3390/s22103872

Academic Editor: Vesa Linnamo

Received: 20 April 2022
Accepted: 17 May 2022
Published: 19 May 2022

1. Introduction

Football is an intermittent sport which alternates different physical actions such as walking, jogging and running at low, medium and high intensities [1]. High-intensity actions (HIAs) and sprints are considered as one of the most important activities in the match and are associated with the most frequent actions in goal situations [2–4].

Global positioning system (GPS) devices are commonly used in several team-sports and help coaches to assess specific movement demands of the players during training sessions and competitions [5]. Many studies of the reliability and validity of GPS in team sports have been made from 2009 [6]. However, there is limited information regarding the validity and reliability of these devices during high-intensity actions [7]. The main findings were that it is a valid method to measure distances at low speed, however, its validity seems to be affected by path linearity and movement intensity [8]. This technology has been demonstrated as a valid and reliable method to analyze the performance across a match, between matches and level of competition, mostly at low intensity actions [6]. More specifically, Coutts and Duffield [7] noticed that GPS has an acceptable level of accuracy and reliability for performance parameters relevant to team sports such as total distance and peak speeds during high-intensity actions, brief and intermittent exercises over a non-linear course.

Performance in HIAs can be greatly different between starters and substitute players [9]. Substitutions during a match are allowed by football rules and usually coaches execute it to change tactics, replacing players who are underperforming or injured, giving playing opportunity to young players or players returning from injury [10].

During the game, substitutes should be able to perform at higher intensities than starters [11]. It has been demonstrated that substitutes covered 25% more HIAs and 63% greater sprinting distances during the final 15 min of a game, compared to starters over the same period [12]. Mean second-half heart rate was significantly higher (84 ± 3 vs. 81 ± 4% maximum heart rate) in substitutes compared with players who completed 90 min [13]. Moreover, substitutes covered greater total distance (TD) and 10% more high-intensity running (HIR), with full-backs being the only position for whom substitutes' HIR did not exceed that of players being replaced [14].

The competition game has been quantified as the most demanding session of the week [15]. An important number of players in the team are not exposed to the total training load (TL) of the game [9]. In this line, Kraemer et al. [16] found that physical fitness performance decreased in non-starter players during the season due to the lack of exposure to competition. For that reason, to try to approach the game's training load experienced by the starters, substitutes need a complementary training session (MD + 1C) [17].

However, it has been demonstrated that MD + 1C training load carried out by non-starters was substantially lower than the magnitude produced by the competition game [9]. Moreover, all training load variables (total distance, energy expenditure, time spent above 90% HR_{max}, accelerations, decelerations and high-intensity running) of this session were significantly lower than regular training on match day −4 (4 days before the competition), which is considered the most demanding session of the week [17]. These sessions are composed of a smaller number of players (~9 vs. ~18 in regular training) and an increase in ball touches, dribbles and duels, but lower physical demands [18]. Thus, even though substitutes performed a complementary training session to compensate the absence of participation during the competition, weekly training load for the starters was largely higher than in the non-starters [9].

During a week with one match, non-starters on average showed a lower total load than starters (up to ~30% less for running and high-speed running) [17]. Moreover, starters accumulated greater (large/very large) perceived training load than the non-starters, the official matches being the source of such differences, as this highlights the general risk of underloading non-starters [9].

For these reasons, for substitutes players it is necessary to have compensatory training sessions with either higher intensity, volume or both, organize additional friendly games or let them play in matches for a lower team [17]. These volume and intensity data need to be collected by GPS with high frequency rate because it provides greater validity to measuring physical parameters such as distance and speed [19]. A 10-Hz GPS demonstrated a lower standard error over a 15 m sprint comparing with 5-Hz and 1-Hz GPS [6].

Previous studies have analyzed the physical demands of non-starters, however none of them have separated the minutes of play of each substitute. Therefore, the purpose of this study was to analyze the differences of physical demands of non-starter players regarding the playing time during the competition and to evaluate the physical demands of the compensatory training for substitute elite players.

2. Materials and Methods

2.1. Sample

Substitute football players' performances from the professional Spanish LaLiga club were analyzed. Three seasons' games were recorded (2016/2017 (second division); 2017/2018 and 2018/2019 (first division)). A total number of 1047 observations were included in this study. Goalkeepers, players who participated for less than 5 min, substitutes who played since the first half and starter players who were substituted were excluded. Players were divided into starters (n = 763) and 3 groups of non-starters (n = 285), depending on the minutes played: 5–15 min (n = 68); 15–30 min (n = 141) and 30–45 (n = 75). All players were informed about the objectives and risks and signed a consent form to participate in this investigation. The study protocol was approved by the Local Ethics Committee (Toledo Hospital, Toledo, Spain) and in accordance with the Code of Ethics of the World Medical Association (Declaration of Helsinki).

2.2. Procedures

Seasons 2016/2017 (second division), 2017/2018 and 2018/2019 (first division) games and the compensatory training the day after the match (MD + 1C) of a professional football team were analyzed with Apex GPS 10 Hz global positioning system (GPS) (STATSports, Newry, N. Ireland), which had been previously validated [20]. This device was situated at the upper back in a vest that was well-adjusted to the body. It provides data on the time of satellite tracking devices and location, and it receives information that determines the signal traffic. The quality of the signal could change depending on the location and environmental obstruction, and data are more accurate together with the addition of triaxial accelerometers, magnetometers and gyroscopes. At least four satellites are required to determine the GPS position trigonometrically. Apex GPS 10 Hz showed distance bias of 1.05 ± 0.87%, 2.3 ± 1.1% and 1.11 ± 0.99% in the 400 m trial, 128.5 m circuit and 20 m trial, respectively, and a V_{peak} bias of 26.5 ± 2.3 km h^{-1}. Data obtained from GPS were downloaded and further analyzed by the STATSports Apex Software (Apex, Brampton, ON, Canada, 10 Hz version 2.0.2.4) [20].

The following variables were collected for all players: the total distance (TD; m) and the distance covered by the player in different high-speed zones: distance in zone 4 (distance covered by the player between 14 and 21 km·h^{-1}; m); distance in zone 5 (distance covered by the player between 21 and 24 km·h^{-1}; m); distance in zone 6 (distance covered by the player above 24 km·h^{-1}; m); peak of maximum velocity (V_{MAX}; km·h^{-1}); average speed (V_{MEAN}; km·h^{-1}); and number of actions above 24 km·h^{-1} (number of sprints; n). All variables were calculated in absolute and relative terms per minute of play.

The compensatory training session the day after the game for substitute players was included in this study (MD + 1C). The objective of this session was to replicate the competition load for those players who had not completed the match [21]. The starting players performed an active recovery session (MD + 1R), but this session was not analyzed because it was not associated with substitute players.

2.3. Statistical Analysis

Data are presented as means ± standard deviations. Kolmogorov–Smirnov distribution test was performed to confirm a normal distribution of the variables and Levene's test to evaluate the homogeneity of the variance. One-way analyses of variance (ANOVA) were performed to analyze the differences between the titular players and the three groups of substitutes. The post hoc analysis was adjusted by the Games–Howell test because the variances were unequal. Secondly, independent-samples t-tests were used to compare the results between divisions for each group of substitutes. Finally, independent-samples t-tests were also used to compare the demands between the titular players of the match and the post training of substitutes. All data were statistically analyzed using SPSS V24.0 for Windows (SPSS Inc., Chicago, IL, USA). The level of significance was set at $p < 0.05$.

3. Results

The results revealed, in relative terms (per minute), differences in distances covered and HIAs between the substitute players and the starters (Table 1; $p < 0.05$). The starting players showed lower total distances covered compared to the substitutes who played

5–15 min (−16.77 m·min^{-1}; CI95%: −30.09 to −3.44; ES: 0.62), 15–30 min (−6.38 m·min^{-1}; CI95%: −9.90 to −2.86; ES: 0.46) and 30–45 min (−3.87 m·min^{-1}; CI95%: −7.12 to −0.62; ES: 0.34). However, maximum velocity reached during the game was higher for those who played the whole game (vs. 5–15 min (+1.91 km·h^{-1}; CI95%: 1.11 to 2.71; ES: 0.93); vs. 15–30 min (+1.04 km·h^{-1}; CI95%: 0.58–1.50; ES: 0.57)). Regarding the minutes played by the substitutes, the analysis of variance showed greater distances covered in zone 4 (14–21 km·h^{-1}) for the players with fewer minutes (5–15 min) compared to the distance accumulated by the players who played 15–30 min (+5.18 m·min^{-1}; CI95%: 0.53–9.82; ES: 0.48), 30–45 min (+7.28 m·min^{-1}; CI95%: 2.62–11.93; ES: 0.75) or starters (+10.32 m·min^{-1}; CI95%: 5.96–14.69; ES: 1.08). Starting players covered fewer distances above 24 km·h^{-1} (vs. 5–15 min (−2.49 m·min^{-1}; CI95%: −3.96 to −1.03; ES: 0.82); vs. 15–30 min (−1.67 m·min^{-1}; CI95%: −2.26 to −1.07; ES: 0.80); vs. 30–45 min (−1.12 m·min^{-1}; CI95%: −1.79 to −0.46; ES: 0.61)) and a lower number of sprints (vs. 5–15 min (−0.11 n·min^{-1}; CI95%: −0.17 to −0.04; ES: 0.79); vs. 15–30 min (-0.09 n·min^{-1}; CI95%: −0.11 to −0.06; ES: 0.86); vs. 30–45 min (−0.06 n·min^{-1}; CI95%: −0.09 to −0.03; ES: 0.65)) compared with substitute players ($p < 0.05$).

Table 1. Distances covered and high-intensity actions by substitutes who play 5–15 min (a), 15–30 min (b), 30–45 min (c) and starters (d).

	5–15 Min (a)	15–30 Min (b)	30–45 Min (c)	Starter (d)
Total Distance (m·min^{-1})	124.78 ± 41.57 ^{d}	114.39 ± 15.22 ^{d}	111.88 ± 10.04 ^{d}	108.01 ± 12.41 ^{a,b,c}
Distance Zone 4 (m·min^{-1})	32.36 ± 13.57 ^{b,c,d}	27.19 ± 7.83 ^{a,d}	25.09 ± 5.79 ^{a,d}	22.04 ± 5.49 ^{a,b,c}
Distance Zone 5 (m·min^{-1})	4.73 ± 3.13 ^{d}	4.44 ± 2.09 ^{d}	3.83 ± 1.65 ^{d}	2.86 ± 1.11 ^{a,b,c}
Distance Zone 6 (m·min^{-1})	5.29 ± 4.56 ^{d}	4.46 ± 2.65 ^{d}	3.92 ± 2.14 ^{d}	2.80 ± 1.52 ^{a,b,c}
Number of sprints (n·min^{-1})	0.26 ± 0.20 ^{d}	0.24 ± 0.12 ^{d}	0.21 ± 0.10 ^{d}	0.15 ± 0.07 ^{a,b,c}
V_{MEAN} (km·h^{-1})	9.59 ± 0.91 ^{d}	9.51 ± 0.67 ^{d}	9.29 ± 0.57 ^{d}	8.79 ± 0.55 ^{a,b,c}
V_{MAX} (km·h^{-1})	28.55 ± 2.46 ^{c,d}	29.41 ± 1.99 ^{d}	30.10 ± 2.04 ^{a}	30.45 ± 1.65 ^{a,b}

Data are mean ± SD; ^{a,b,c,d} Significant differences between groups ($p < 0.05$).

Regarding the category, non-starters who played 30–45 min in the second division reported higher peak of maximum velocity than in the first division (+1.18 km·h^{-1}; CI95%: 0.23–2.13; ES: 0.61; $p < 0.05$). No difference was observed between league standards in any of the other variables ($p > 0.05$; Figure 1).

The compensatory training session (MD + 1C) for substitutes reported significant lower values in all variables compared with starters who played the complete game, especially in sprint distance (−230.48 m; CI95%: −244.08 to −216.88; ES: 2.29; $p < 0.05$). Absolute values revealed a significant decrement of physical demands in non-starters with fewer minutes of play during the game ($p < 0.05$; Table 2).

Table 2. Absolute distances covered and high-intensity actions during compensatory training for substitutes (MD + 1C), for starters on match day and non-starters depending on the minutes of play.

	MD + 1C	MD Starters	MD 5–15 min NS	MD 15–30 min NS	MD 30–45 min NS
Total Distance (m)	4654.89 ± 919.59 *	10,171.23 ± 955.22	1310.50 ± 463.35 *	2495.90 ± 571.02 *	3841.76 ± 468.72 *
Distance Zone 4 (m)	447.44 ± 330.92 *	2075.72 ± 508.68	339.10 ± 153.63 *	589.95 ± 188.05 *	857.94 ± 190.93 *
Distance Zone 5 (m)	71.33 ± 63.4 *	269.60 ± 103.05	49.00 ± 31.50 *	95.56 ± 45.47 *	130.55 ± 54.50 *
Distance Zone 6 (m)	33.41 ± 57.64 *	263.90 ± 143.70	53.47 ± 43.24 *	96.19 ± 56.97 *	134.39 ± 72.84 *
Number of sprints (*n*)	3.96 ± 3.74 *	14.25 ± 7.08	2.62 ± 1.90 *	5.07 ± 2.61 *	7.17 ± 3.47 *
V_{MEAN} (km·h^{-1})	3.84 ± 0.59 *	8.79 ± 0.55	9.59 ± 0.91 *	9.51 ± 0.67 *	9.29 ± 0.57 *
V_{MAX} (km·h^{-1})	26.48 ± 3.23 *	30.45 ± 1.65	28.55 ± 2.46 *	29.41 ± 1.99 *	30.10 ± 2.04 *

Data are mean ± SD; MD: match day; NS: non-starters; V_{MEAN}: mean speed; V_{MAX}: maximum speed; * significant differences with MD starters ($p < 0.05$).

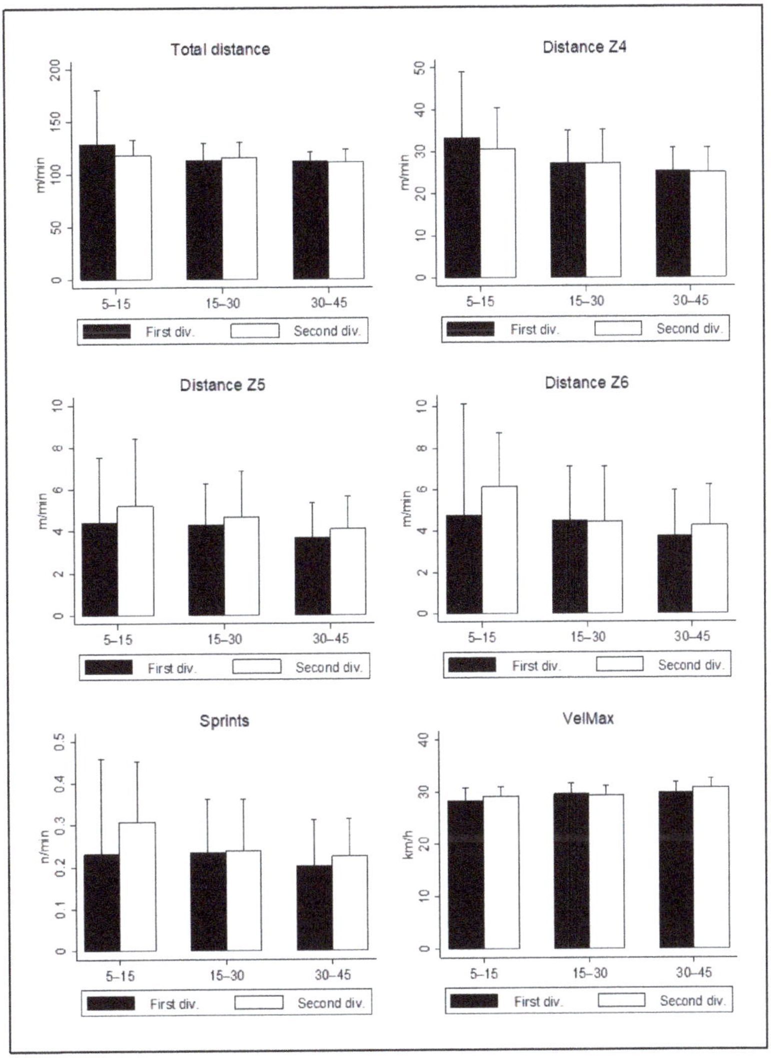

Figure 1. Distances covered and high-intensity actions by substitutes who play 5–15 min, 15–30 min, 30–45 min. Differences between the 1st and 2nd divisions.

4. Discussion

In the present study, 10 Hz GPS devices (Apex GPS, STATSports, Newry, N. Ireland), have been used to collect all performance data for substitutes during games and MD + 1C, because these units are capable of measuring the smallest changes in acceleration and deceleration, whereas, lower frequency rate units are unable [22,23]. The substitutes performed, in relative terms, higher physical actions compared with starters, and those who played the shortest period of time during the game (5–15 min) gave the highest physical performance compared with the other substitutes who played for longer times (15–30 and 30–45 min). Thus, the less time the non-starters play in the game, the higher relative physical performance they reach. This information will be useful to plan more individualized training sessions adapted to each substitute according to the time of participation during the match, and avoid the same stimulus for all substitutes, therefore, incorrect training load for each player.

Fewer data exist regarding the responses of substitutes entering the field of play [10]. It is difficult to compare our data among substitutes' time-play with the literature because, to our knowledge, no studies have investigated non-starters' physical demands regarding the time-play during the competition. Although the time-play of substitutes has not been considered, Bradley et al. [14] mentioned that the players introduced must be immediately able to perform at equivalent or higher work rates than either the players being replaced, others remaining on the pitch or both. In accordance with our investigation, several studies noticed that elite substitutes who had been introduced during the second half covered 25% more HIR and 63% greater sprinting distances during the final 15 min of a game, compared to whole-match players over the same period [23–25]. In contrast to our findings, which demonstrated that the shortest time on the pitch is related to a higher physical performance in relative terms, preliminary investigations have indicated an inability of players introduced as substitutes to exceed the running performance that they typically adopt during the first half of matches that they start [24]. Thus, non-starters perform greater physical actions when they are introduced in the game compared with starters.

It is important to notice that GPS devices used in this study have been demonstrated as a valid method to measure distances at low and moderate but not high intensity [19].

Another aim of this study was to analyze the differences between league standard in HIAs of non-starters. No differences were found between first and second divisions regarding physical performance of non-starters. These results are in contrast with several researches that have shown that players at a higher standard of play perform more high-intensity running than peers at lower standards [24,26,27]. In the elite Italian League, players performed 28% more high-intensity running than sub-elite Danish League peers [24]. However, the data were captured from two separate European Leagues of vastly different standards. Similarly, Ingebrigtsen et al. [27] reported that distance covered in high-intensity running was 30–40% greater in players in top versus middle- and bottom-ranking Danish teams. However, in contrast with these findings, Di Salvo et al. [28] showed that Championship players covered greater distances in jogging, running, high-speed running and sprinting than Premier League players. Another similar investigation found that players in League 1 and the Championship performed more high-intensity running than those in the Premier League. Players also covered more high-intensity running when moving down from the Premier League to the Championship but not when players moved up standards [29]. All these studies focused on starters' performance and are compared with our findings that refer to non-starter players because, to our knowledge, no investigation has analyzed the differences in physical performance of substitute players across two standards of elite football within a single country.

The present study also evaluated the physical demands of the compensatory training (MD + 1C) for substitutes. Players with reduced game time will require a training session that replicates competition loads, whereas those players completing the game will require a recovery session instead (MD + 1R) [17]. Our results found that MD + 1C load is lower than competition load. Accordingly, Stevens et al. [17] noted that non-starters training on the day after the match showed significantly lower values than starters on total distance ($-17 \pm 5\%$), energy expenditure ($-18 \pm 5\%$), time spent above 90%HRmax ($-52 \pm 35\%$), running ($-55 \pm 13\%$), accelerations ($-42 \pm 12\%$), decelerations ($-46 \pm 15\%$) and high power ($-30 \pm 9\%$), contributing on average to a lower estimated total load. In contrast, Martín García et al. [21] showed that MD + 1 exceed 50% of match play values, and these included total distance covered (53%), average metabolic power (69%), accelerations (86%) and decelerations (80%), and could be an ideal day to compensate for the reduced competition load in players with limited game time. Moreover, the present study found that the accumulation of match game load and MD + 1C training session for non-starters are lower than full game load for starters. In line with this, the match was the most demanding session of the week only for the starters, contributing to considerably lower values for substitutes' training [17], physical fitness performance decrements [19] and a general risk of under-loading non-starters having injuries [30]. Thus, competition time is the main source of differences between starters and non-starters in accumulated training load [9]. These results are important for coaches in order to stimulate substitutes players with additional training load until reaching the full-game accumulation load. Moreover, the game time played by each substitute monitored will help to individualized training load in different groups according this time of participation during the competition.

In absolute terms, non-starters' load accumulation depends on the game time during the competition and it is difficult to compare with other studies because no studies have separated time-play periods of substitutes. Altogether, there is clearly a challenge to sufficiently load non-starters within the specific and individual context of each player. Further research is needed to refine training prescription of compensatory training sessions for non-starters to ensure their readiness for competition, considering specific game time played.

5. Conclusions

The present study demonstrates that substitutes players performed higher physical actions at high-intensity (total distance, accelerations, number of sprints, V_{max}, V_{mean}) per minute compared with starters. Moreover, substitutes who played the shortest period of time during the game (5–15 min) produced the highest physical performance compared with the other substitutes who played more time (15–30 and 30–45 min). Therefore, substitutes' game loads depend on the time of playing. Regarding the league standard, the physical performance of substitutes was not significantly different between first and second divisions. Additionally, compensatory training (MD + 1C) accumulated training load has been demonstrated as lower comparing with the full game, showing that substitutes are at risk of being undertrained, have a reduced physical performance and increased risk of injury. These results increase the knowledge for coaches to plan training sessions according to the time-play of each substitute and guarantee a more individualized training load which will aid to decrease injuries and optimize performance during the competition.

Author Contributions: All authors contributed an important and collaborative participation in this study. Writing–original draft preparation, G.C.-P. and J.S.-S.; data curation, J.G.-U. and A.H.-M.; validation, J.G.-U.; conceptualization, A.H.-M.; formal analysis, J.S.-S. and L.G.; writing–review and editing, G.C.-P.; methodology, V.P.-H.; investigation, J.L.F.; supervision, J.S.-S. and G.C.-P. All authors have read and agreed to the published version of the manuscript.

Funding: This research received no external funding.

Institutional Review Board Statement: The study was conducted in accordance with the Declaration of Helsinki and approved by the Institutional Review Board (or Ethics Committee) of Local Ethics Committee (Toledo Hospital, Toledo, Spain) (Ref.: 489/24022020 24/02/2020).

Informed Consent Statement: Informed consent was obtained from all subjects involved in the study.

Data Availability Statement: Not applicable.

Acknowledgments: The authors would like to thank the football sport schools for their support and collaboration, as well as all the football players who contributed in this research. A.H.-M. acknowledges the Spanish Ministry of Science, Innovation and Universities for funding the development of his PhD (Grant Number: FPU18/03222). J.G.-U. acknowledges "Fondo Europeo de Desarrollo Regional, Programa Operativo de la Región de Castilla-La Mancha" (2018/11744) for funding the development of his research.

Conflicts of Interest: The authors declare no conflict of interest.

References

1. Moura, F.A.; Marche, A.L.; Caetano, F.G.; da Silva Torres, R.; Martins, L.E.B.; Cunha, S.A. Analysis of high-intensity efforts in Brazilian professional soccer players. *Hum. Mov. Spec. Issue* **2017**, *18*, 55–62. [CrossRef]
2. Di Salvo, V.; Gregson, W.; Atkinson, G.; Tordoff, P.; Drust, B. Analysis of high intensity activity in Premier League soccer. *Int. J. Sports Med.* **2009**, *30*, 205–212. [CrossRef] [PubMed]
3. Faude, O.; Koch, T.; Meyer, T. Straight sprinting is the most frequent action in goal situations in professional football. *J. Sports Sci.* **2012**, *30*, 625–631. [CrossRef] [PubMed]
4. Bradley, P.S.; Lago-Peñas, C.; Rey, E.; Sampaio, J. The influence of situational variables on ball possession in the English premier league. *J. Sports Sci.* **2014**, *32*, 1867–1873. [CrossRef]
5. Gray, A.J.; Jenkins, D.; Andrews, M.H.; Taaffe, D.R.; Glover, M.L. Validity and reliability of GPS for measuring distance travelled in field-based team sports. *J. Sports Sci.* **2010**, *28*, 1319–1325. [CrossRef]
6. Aughey, J.R. Applications of GPS Technologies to Field Sports. *Int. J. Sports Physiol. Perform.* **2011**, *6*, 295–310. [CrossRef]
7. Coutts, A.J.; Duffield, R. Validity and reliability of GPS devices for measuring movement demands of team sports. *J. Sci. Med. Sports* **2008**, *13*, 133–155. [CrossRef]
8. Witte, T.H.; Wilson, A.M. Accuracy of non-differential GPS for the determination of speed over ground. *J. Biomech.* **2004**, *37*, 1891–1898. [CrossRef]
9. Los Arcos, A.; Mendez-Villanueva, A.; Martínez-Santos, R. In-season training periodization of professional soccer players. *Biol. Sport* **2017**, *34*, 149–155. [CrossRef]
10. Hills, S.P.; Barwood, M.J.; Radcliffe, J.N.; Cooke, C.B.; Kilduff, L.P.; Cook, C.J.; Russell, M. Profiling the responses of soccer substitutes: A review of current literature. *Sports Med.* **2018**, *48*, 2255–2269. [CrossRef]
11. Waldron, M.; Highton, J. Fatigue and pacing in high-intensity intermittent team sport: An update. *Sports Med.* **2014**, *44*, 1645–1658. [CrossRef] [PubMed]
12. Bradley, P.S.; Noakes, T.D. Match running performance fluctuations in elite soccer: Indicative of fatigue, pacing or situational influences? *J. Sports Sci.* **2013**, *31*, 1627–1638. [CrossRef] [PubMed]
13. Coelho, D.B.; Coelho, L.G.; Morandi, R.F.; Ferreira-Júnior, J.B.; Bouzas Marins, J.C.; Prado, L.S.; Danusa Dias, S.; Silami-Garcia, E. Effect of player substitutions on the intensity of second-half soccer match play. *Rev. Bras. Cineantropometria* **2012**, *14*, 183–191.
14. Bradley, P.S.; Lago-Peñas, C.; Rey, E. Evaluation of the match performances of substitution players in elite soccer. *Int. J. Sports Physiol. Perform.* **2014**, *9*, 415–424. [CrossRef] [PubMed]
15. Alexiou, H.; Coutts, A.J. A comparison of methods used for quantifying internal training load in women soccer players. *Int. J. Sports Physiol. Perform.* **2008**, *3*, 320–330. [CrossRef] [PubMed]
16. Kraemer, W.J.; French, D.N.; Paxton, N.J.; Häkkinen, K.; Volek, J.S.; Sebastianelli, W.J.; Putukian, M.; Newton, R.U.; Rubin, M.R.; Gómez, A.L.; et al. Changes in exercise performance and hormonal concentrations over a big ten soccer season in starters and nonstarters. *J. Strength Cond. Res.* **2004**, *18*, 121–128.
17. Stevens, T.G.A.; de Ruiter, C.J.; Twisk, J.W.R.; Savelsbergh, G.J.P.; Beek, P.J. Quantification of in-season training load relative to match load in professional Dutch Eredivisie football players. *Sci. Med. Foot* **2017**, *1*, 117–125. [CrossRef]
18. Owen, A.L.; Wong, D.P.; Paul, D.; Dellal, A. Physical and technical comparisons between various-sided games within professional soccer. *Int. J. Sports Med.* **2014**, *35*, 286–292. [CrossRef]
19. Cummins, C.; Orr, R.; O'Connor, H.; West, C. Global Positioning Systems (GPS) and Micrtechnology Sensors in Team Sports: A systematic Review. *Sports Med.* **2013**, *43*, 1025–1042. [CrossRef]
20. Beato, M.; Coratella, G.; Stiff, A.; Dello Iacono, A. The validity and between-unit variability of GNSS units (STATSports Apex 10 and 18 Hz) for measuring distance and peak speed in team sports. *Front. Physiol.* **2018**, *9*, 1288. [CrossRef]
21. Martín-García, A.; Gómez Díaz, A.; Bradley, P.S.; Morera, F.; Casamichana, D. Quantification of a professional football team's external load using a microcycle structure. *J. Strength Cond. Res.* **2018**, *32*, 3511–3518. [CrossRef] [PubMed]
22. Hernandez-Martin, A.; Sanchez-Sanchez, J.; Felipe, J.L.; Manzano-Carrasco, S.; Majano, C.; Gallardo, L.; Garcia-Unanue, J. Physical Demands of U10 Players in a 7-a-Side Soccer Tournament Depending on the Playing Position and Level of Opponents in Consecutive Matches Using Global Positioning Systems (GPS). *Sensors* **2020**, *20*, 6968. [CrossRef] [PubMed]
23. Varley, M.; Aughey, R. Validity and reliability of GPS for measuring instantaneous velocity during acceleration, deceleration, and constant motion. *J. Sports Sci.* **2012**, *30*, 121–127. [CrossRef] [PubMed]
24. Carling, C.; Espié, V.; Le Gall, F.; Bloomfield, J.; Jullien, H. Work-rate of substitutes in elite soccer: A preliminary study. *J. Sci. Med. Sport* **2010**, *13*, 253–255. [CrossRef]
25. Mohr, M.; Krustrup, P.; Bangsbo, J. Match performance of high-standard soccer players with special reference to development of fatigue. *J. Sports Sci.* **2003**, *21*, 519–528. [CrossRef]
26. Andersson, H.A.; Randers, M.B.; Heiner-Moller, A.; Krustrup, P.; Mohr, M. Elite female soccer players perform more high-intensity running when playing in international games compared with domestic league games. *J. Strength Cond. Res.* **2010**, *24*, 912–919. [CrossRef]
27. Ingebrigtsen, J.; Bendiksen, M.; Randers, M.B.; Castagna, C.; Krustrup, P.; Holtermann, A. Yo-Yo IR2 testing of elite and sub-elite soccer players: Performance, heart rate response and correlations to other interval tests. *J. Sports Sci.* **2012**, *30*, 1337–1345. [CrossRef]

28. Di Salvo, V.; Pigozzi, F.; Gonzalez-Haro, C.; Laughlin, M.S.; De Witt, J.K. Match running performance comparisons in top English soccer leagues. *Int. J. Sports Med.* **2013**, *34*, 526–532.
29. Bradley, P.S.; Carling, C.; Gomez Diaz, A.; Hood, P.; Barnes, C.; Ade, J.; Boddy, M.; Krustrup, P.; Mohr, M. Match performance and physical capacity of players in the top three competitive standards of English professional soccer. *Hum. Mov. Sci.* **2013**, *32*, 808–821. [CrossRef]
30. Gabbett, T.J. The training-injury prevention paradox: Should athletes be training smarter and harder? *Br. J. Sports Med.* **2016**, *50*, 273–280. [CrossRef]

sensors

MDPI

Article

Propulsion Calculated by Force and Displacement of Center of Mass in Treadmill Cross-Country Skiing

Shuang Zhao [1], Olli Ohtonen [1], Keijo Ruotsalainen [1], Lauri Kettunen [2], Stefan Lindinger [3], Caroline Göpfert [4] and Vesa Linnamo [1,*]

1 Faculty of Sport and Health Sciences, University of Jyväskylä, 40014 Jyväskylä, Finland; zhaoshuangzs@hotmail.com (S.Z.); olli.ohtonen@jyu.fi (O.O.); keijo.s.ruotsalainen@gmail.com (K.R.)
2 Faculty of Information Technology, University of Jyväskylä, 40014 Jyväskylä, Finland; lauri.y.o.kettunen@jyu.fi
3 Center of Health and Performance (CHP), Department of Food and Nutrition and Sport Science, University of Gothenburg, 40530 Göteborg, Sweden; stefan.lindinger@gu.se
4 Department of Sport Science and Kinesiology, University of Salzburg, 5400 Salzburg, Austria; caro.goepfert@gmx.de
* Correspondence: vesa.linnamo@jyu.fi

Abstract: This study evaluated two approaches for estimating the total propulsive force on a skier's center of mass (COM) with double-poling (DP) and V2-skating (V2) skiing techniques. We also assessed the accuracy and the stability of each approach by changing the speed and the incline of the treadmill. A total of 10 cross-country skiers participated in this study. Force measurement bindings, pole force sensors, and an eight-camera Vicon system were used for data collection. The coefficient of multiple correlation (CMC) was calculated to evaluate the similarity between the force curves. Mean absolute force differences between the estimated values and the reference value were computed to evaluate the accuracy of each approach. In both DP and V2 techniques, the force–time curves of the forward component of the translational force were similar to the reference value (CMC: 0.832–0.936). The similarity between the force and time curves of the forward component of the ground reaction force (GRF) and the reference value was, however, greater (CMC: 0.879–0.955). Both approaches can estimate the trend of the force–time curve of the propulsive force properly. An approach by calculating the forward component of GRF is a more appropriate method due to a better accuracy.

Keywords: propulsive force; V2-skating skiing technique; double-poling skiing technique

Citation: Zhao, S.; Ohtonen, O.; Ruotsalainen, K.; Kettunen, L.; Lindinger, S.; Göpfert, C.; Linnamo, V. Propulsion Calculated by Force and Displacement of Center of Mass in Treadmill Cross-Country Skiing. *Sensors* **2022**, *22*, 2777. https://doi.org/10.3390/s22072777

Academic Editor: Christian Peham

Received: 17 January 2022
Accepted: 2 April 2022
Published: 5 April 2022

1. Introduction

Forces acting on a skier's center of mass (COM) in a forward direction are propulsive forces, which are the primary mechanical determinants of an cross-country (XC) skier's performance [1]. The position of skier's COM can be obtained by using the marker-based motion capture system with a segmental method [2]. Thus, forces acting on a skier's COM can be obtained by multiplying COM acceleration with the total mass of the skier, and this will indicate how athletes overcome resistive forces. However, the contribution of single pole and leg thrusts could not be revealed. Therefore, it is essential to compute forces acting on the COM from the ground reaction forces (GRFs) generated from skis and poles, separately.

Except for estimating the propulsive force with the forward acceleration of COM and the total mass, other approaches have been developed. One approach is to estimate the propulsive force as the forward-directed horizontal component of the three-dimensional (3D) GRFs from both skis and poles that act on a skier (F_{net}) [3]. The roller skis [3–5], skis [6,7], and poles [3,6,7] equipped with force sensors have been used to measure the forces generated from skis and poles. Combined with the pole angle, ski angle, ski-edging angle, and the incline of the track or the treadmill, the propulsive force from skis and poles can be specified [1,3,8]. Therefore, questions related to the propulsive force, including the contribution of skis and poles in different techniques [3,9,10], and the comparison of different techniques [11,12] have been addressed. Another approach, demonstrated by Göpfert et al. [6], is to estimate the propulsive force with the forward component of translational force (F_{pro}). The translational force was modeled as the component of the 3D resultant GRFs that acts in the direction from the point of force application (PFA) to a skier's COM [6], and calculated by projecting the GRFs to the line defined by the COM and PFA.

The propulsive forces obtained with the two mentioned approaches (F_{net} and F_{pro}) have been compared to the propulsive force calculated with COM acceleration from a motion analysis system (F) in [6]. As using the segmental method has been shown to be suitable for estimating the position of the COM in sports [2], F was considered as the reference value. The results indicated that the force–time curves of F_{net} and F_{pro} all showed high similarity when compared to the force–time curves of F during the leg skating push-offs on snow. F_{net} overestimated F, and F_{pro} was found to be a more appropriate approach to estimate F during leg skating push-offs [6]. However, whether F_{net} and F_{pro} could be used to estimate F, and which one is more accurate in other techniques, are still unknow. As XC skiing is a sport whose competition and training are normally performed on varying track topography and speed, whether F_{net} and F_{pro} could work steadily when estimating F at different terrain with different speeds need further investigation.

Therefore, the first aim of the present study was to obtain the force–time curves of F_{net} and F_{pro} with different skiing techniques and evaluate which can estimate the force–time curves of F better. As the use and importance of double poling (DP) and V2 skating (V2) as main techniques in XC skiing have increased for the past few years [12–14], DP and V2 skating techniques will be performed in this study. The second aim is to investigate which approach is more accurate when estimating F. Another aim is to explore the stability of the approaches to calculate F_{net} and F_{pro} by changing the speed and incline of the treadmill. We hypothesized that the force–time curves of F_{net} and F_{pro} all give comparable shape with F in both techniques [6]. We also hypothesized that F_{net} would give a considerable overestimation, and F_{pro} would be more accurate than F_{net}, when estimating F in both DP and V2 techniques [6]. We further hypothesized that the approaches to calculate the F_{net} and F_{pro} would not be affected by the speed and incline of the treadmill in both techniques.

2. Materials and Methods

2.1. Participants

A total of 10 experienced male skiers (age: 29.4 ± 7.9 years; height: 181.4 ± 5.7 cm; weight: 77.9 ± 8.9 kg) who were familiar with treadmill roller skiing volunteered to participate in this study. The experimental protocol and all methods used in this study were approved by the Ethics Committee of the University of Jyväskylä. All participants provided written informed consent before the measurement and were free to withdraw from the experiments at any point.

2.2. Protocol

The anthropometric parameters needed for motion analysis (e.g., bilateral leg length, knee width, ankle width, shoulder offset, elbow width, and hand thickness) were measured first, and passive reflective markers were attached to the participants and equipment. Once the preparations were made, participants completed a 10–15 min warm-up roller skiing on the treadmill. Next, calibration was performed with the skier in a standing position and the treadmill at a 0° incline. Participants then performed the DP technique at five speeds (13, 15, 17, 19, and 21 km/h) on a 2° incline. The comfortable pole length for the DP technique was 1.56 ± 0.06 m. After the trials with varying speeds at a 2° incline, the DP technique was performed at three inclines (3°, 4°, and 5°) with a speed of 10 km/h. There was a 1 min rest between each speed and incline. When participants finished performing the DP technique, the pole length was adjusted to a comfortable length for the V2 skating technique (which was 1.63 ± 0.03 m in this study). The participants were given a short rest period while adjusting the pole length. The participants then performed the V2 technique on the treadmill. The protocol for speed and incline change was the same as during the DP test.

2.3. Data Collection

An eight-video-camera motion capture system (Vicon, Oxford, UK) and NEXUS 2.8.1 software were used to collect and record the 3D trajectories of reflective markers at a sampling rate of 150 Hz. The global coordinate system (GCS) was defined using the right-hand rule when the incline of the treadmill was 0°. The *X*-axis of the GCS was defined as the direction from side to side across the treadmill. The *Y*-axis of the GCS was the longitudinal axis of the treadmill. The *Z*-axis of the GCS was perpendicular to the ground, pointing upward. The GCS was calibrated according to Vicon's specifications. A total of 58 passive reflective markers were used in this current study: 43 passive reflective markers were attached to the participants' bodies, and 15 markers were attached to the equipment, including both skis (3 each), both poles (3 each), and the treadmill (3). Anthropometric measurements and the placement of markers on the participants' bodies were conducted according to the XC model [6] used in previous studies. Measurements were performed on a motorized treadmill with a belt surface of 2.7 m wide and 3.5 m long (Rodby Innovation AB, Vänge, Sweden). The same pair of roller skis were used for both techniques (Marwe, SKATING 620 XC, wheel no. 0), with a resistance friction coefficient of μ = 0.025 measured before the measurement (Appendix A).

Two custom-made pole force sensors (VTT MIKES, Technical Research Centre of Finland Ltd., Kajaani, Finland, Figure 1a) were used to measure the axial GRF from the poles. Two custom-made two-dimensional (2D) force measurement bindings (Neuromuscular Research Centre, University of Jyväskylä, Jyväskylä, Finland, Figure 1b) [15] were mounted on roller skis to measure the leg forces generated from roller skis. Both pole and leg forces were collected synchronously with the Coachtech online measurement and feedback system (Neuromuscular Research Centre, University of Jyväskylä, Jyväskylä, Finland) at a sample rate of 400 Hz. The force measurement bindings measured the vertical (F_{skiz}) and mediolateral (F_{skix}) forces and were calibrated before the measurement [15]. A trigger signal was sent from the Coachtech [16] to the motion capture system to mark the start of the force capture. The nodes for the pole force sensors and force measurement bindings were used to supply power and transmit data.

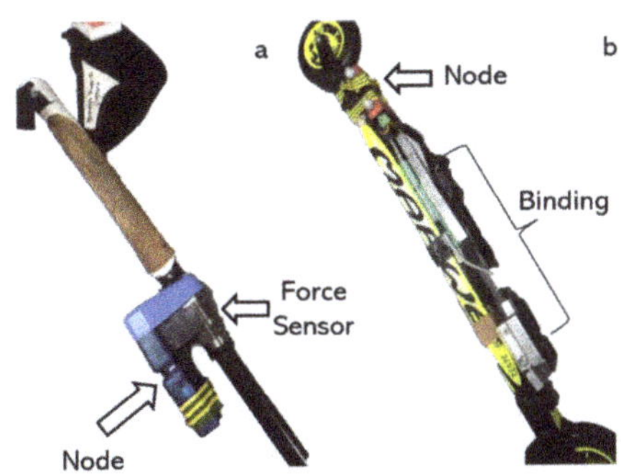

Figure 1. Equipment used in this study: (**a**) Pole force measurement sensor. (**b**) Force measurement binding.

2.4. Data Reduction

Marker labeling and COM calculations were performed using NEXUS 2.8.1 software. The raw 3D trajectories of all reflective markers and the acceleration of COM were low-pass filtered (fourth-order, zero-lag, and Butterworth filter) with a cutoff frequency of 11.3 Hz [17]. The XC model [6], which contained the head, thorax, abdomen and pelvis, upper arms, hands, thighs, shanks, feet, skis, and poles, was used to calculate the whole-body COM. The marker placement on the subject and geometric model for the XC model is shown in Figure 2. The segmental anthropometric data were taken from Dempster's study as described in Selbie et al. [18]. Force data were low-pass filtered (eighth-order, zero-lag, and Butterworth filter) with a cutoff frequency of 15 Hz [19]. Data filtering and parameter calculations were performed using MATLAB R2018a (MathWorks, Natick, MA, USA).

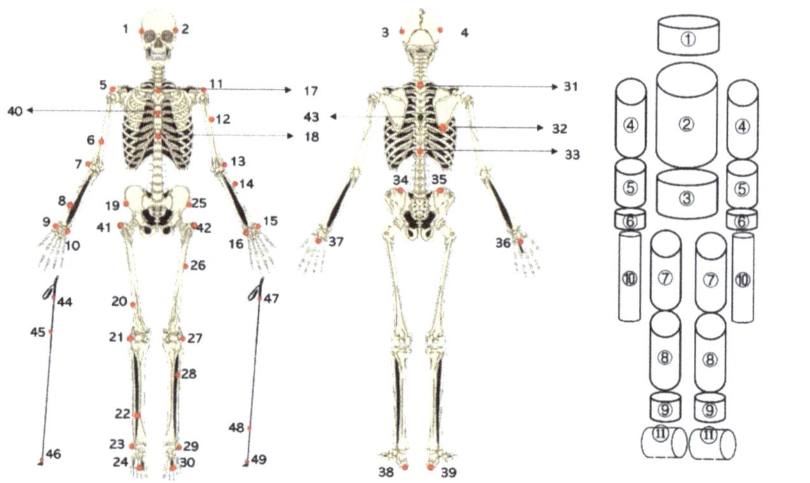

Figure 2. Marker placement on the subject and geometric model for segments in the XC model. The numbers 1–49 represent the placement of reflective markers on subjects and poles. The displacement of reflective markers on roller skis is shown in Figure 3. The numbers 1–39 are the markers used in the plug-in-gait (PIG) model. 1–43 are the markers used in the XC model [6] on one subject. The numbers 44–49 are the markers on the poles. ①–⑪ represent the head, thorax, abdomen and pelvis, upper arm, forearm, hand, thigh, shank, foot, pole, and roller ski, respectively.

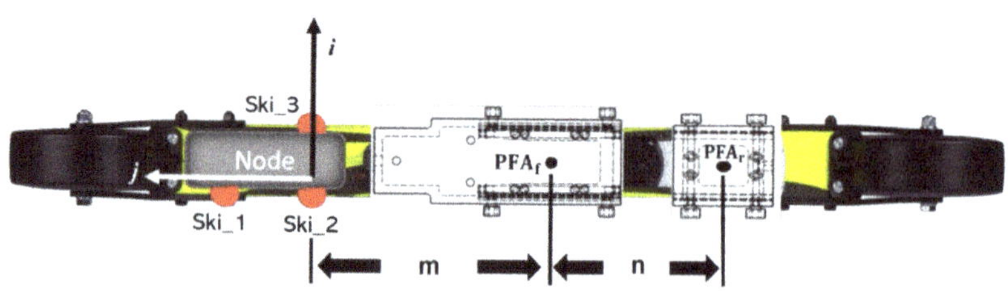

Figure 3. Displacement of markers, PFAs, and the definition of FCS ($\vec{i}$, $\vec{j}$, $\vec{k}$). Three markers (Ski_1, Ski_2, and Ski_3) were attached to the side of the node. The node for power supply and data transmission was attached to the front part of the roller ski. The surface defined by the markers was parallel to the roller ski surface. $\vec{i}$ was defined by Ski_3 and Ski_2. Another unit vector ($\vec{r}$) located on the surface of the roller ski was defined by Ski_1 and Ski_2. The surface norm, which was the $\vec{k}$ of FCS, was the cross product of $\vec{i}$ and $\vec{r}$. The last unit vector $\vec{j}$ was computed by using the right-hand rule with $\vec{k}$ and $\vec{i}$. The PFA_f and PFA_r were the points of force application of the front and rear sensors, respectively. The distance between Ski_2 and PFA_f was m, and the distance between PFA_f and PFA_r was n.

2.4.1. Transforming the Forces Measured from the Force Sensor into the GCS and the PFA

The forces generated from the roller ski force coordinate system (FCS) were transformed into the GCS. The unit vector of each axis of FCS ($\vec{i}$, $\vec{j}$, $\vec{k}$) was identified by markers on the roller ski (Figure 3). The transformation from the roller ski system to the GCS is given by

$$\begin{bmatrix} F_x \\ F_y \\ F_z \end{bmatrix} = R' \begin{bmatrix} F_{skix} \\ 0 \\ F_{skiz} \end{bmatrix} \tag{1}$$

where F_x, F_y, and F_z are the components of forces generated from legs ($\overrightarrow{F_r}$) in the GCS. R' [18] is the rotation matrix from FCS to GCS.

The PFA is needed for calculating the translational force introduced by Göpfert et al. [6]. The displacement of the PFA along the binding (PFA_{ski}) was calculated from the force distribution between the front and rear sensors of the binding [20] over time. The PFA for each part of the binding (PFA_f and PFA_r) was defined as the center of each sensor. The distance between the marker Ski_2 and PFA_f (m), and the distance between PFA_f and PFA_r (n) were measured before the measurement. The displacements of PFA_f and PFA_r in GCS were obtained by moving the midpoint of Ski_2 and Ski_3 along the opposite direction of $\vec{j}$. The moving distances were m and $m + n$, respectively. The mediolateral sway of PFA_{ski} on ski binding was not considered in this study. Thus, the PFA_{ski} moved between the PFA_f and PFA_r (Figure 3).

The measured axial pole forces ($\overrightarrow{F_p}$) were considered the GRFs acting along the pole from the tip to the top of the pole and expressed that way in the GCS. The magnitude of $\overrightarrow{F_p}$ was collected using a pole force sensor. The direction of $\overrightarrow{F_p}$ was defined using the reflective markers that were attached to the pole. The PFA of poles (PFA_p) was defined as the intersection of the plane of the treadmill and the long axis of the pole. The plane of the treadmill was defined using the three markers attached to the treadmill.

2.4.2. The Reference Force, the Total Resultant Force, and the Translational Force

As using the segmental method has been shown to be suitable for estimating the position of the COM in sports [2], forces calculated by COM acceleration ($\vec{a}$) multiplied the total mass of the subject, and the equipment was the reference force ($\vec{F}$) in this study.

One approach to estimate forces acting on skier's COM is to calculate the total resultant force ($\overrightarrow{F_{net}}$) without considering the position of COM. $\overrightarrow{F_{net}}$ is calculated as

$$\overrightarrow{F_{net}} = \overrightarrow{F_r} + \overrightarrow{F_p} + \overrightarrow{F_{friction}} + \vec{G} \tag{2}$$

where $\vec{G}$ is the gravitational force of each participant and all the equipment. $\overrightarrow{F_{friction}}$ is the frictional force between the roller ski and the treadmill, which was directed along the path of the ski motion, and the magnitude was computed by multiplying μ with F_{skiz}.

Another approach to estimate forces acting on skier's COM is to calculate the total translational force (the mechanical principle of translational force, see Appendix B). The translational force is the share of the resultant GRF acting in the direction from PFA to COM. The translational force from skis ($\overrightarrow{F_{tS}}$, Figure 4) is the share of ski GRF ($\overrightarrow{F_r}$) acting in the direction defined from PFA_{ski} to COM and is calculated from

$$\overrightarrow{F_{tS}} = (\overrightarrow{F_r} \bullet \vec{u})\vec{u} \tag{3}$$

where $\vec{u}$ is the unit vector determined from PFA_{ski} to COM. The translational force from poles ($\overrightarrow{F_{tP}}$) is calculated from

$$\overrightarrow{F_{tP}} = (\overrightarrow{F_p} \bullet \vec{v})\vec{v} \tag{4}$$

where $\vec{v}$ is the unit vector determined from PFA_p to COM. The total translational force ($\overrightarrow{F_{pro}}$) is the sum of the translational force from the legs, poles, and the resistance. Thus, $\overrightarrow{F_{pro}}$ can be computed as

$$\overrightarrow{F_{pro}} = \overrightarrow{F_{tS}} + \overrightarrow{F_{tP}} + \overrightarrow{F_{friction}} + \vec{G} \tag{5}$$

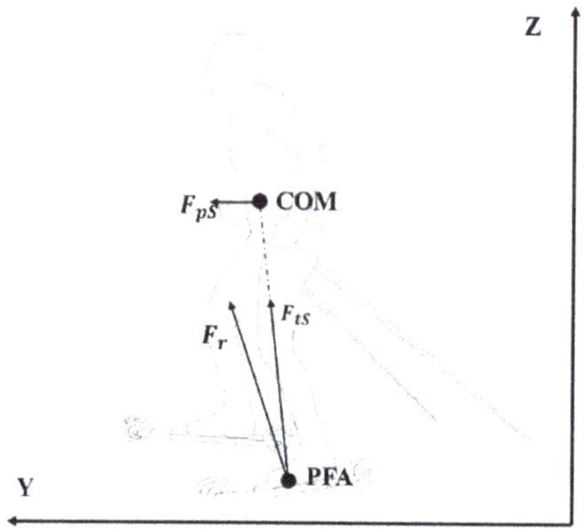

Figure 4. Diagram of force decomposition from skis. F_r is the resultant force generated from legs. F_{tS} is the translational component, which went through the COM. F_{pS} represents the propulsion generated from legs in the forward direction.

As forces acting on skier's center of mass (COM) in forward direction are the propulsive forces, the Y component of $\vec{F}$, $\overrightarrow{F_{net}}$, and $\overrightarrow{F_{pro}}$ (F, F_{net}, and F_{pro}) was compared and analyzed in the present study.

2.4.3. Cycle Definition and Analyzed Parameters

A total of 10 consecutive poling phases for each DP technique trial and 10 consecutive kicking phases (5 left ski kicking and 5 right ski kicking) for each V2 technique trial were analyzed. The poling phase was defined as the period from the start of the pole ground contact to the end of the pole ground contact (Figure 5a). The kicking phase was defined as the ski force minima until the end of ground contact [7] (Figure 5b). The forces of skis and poles from both the left and right sides were included while calculating the total propulsion in both techniques.

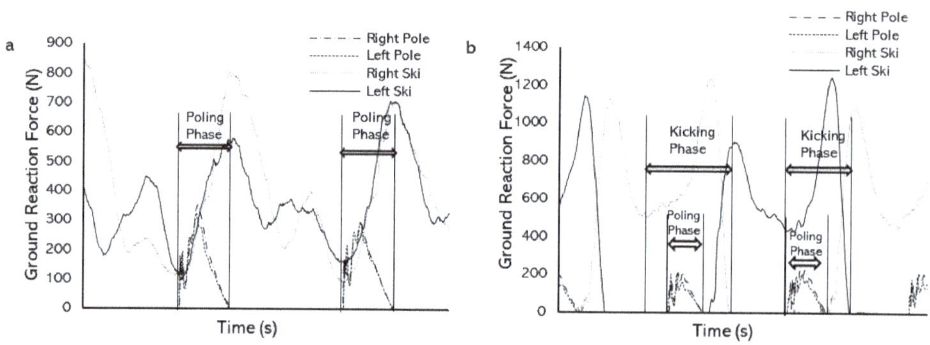

Figure 5. Definition of the force producing phase of DP and V2 techniques: (**a**) GRFs from skis and poles in the DP technique and the definition of the poling phase. (**b**) GRFs from skis and poles in the V2 technique and the definition of the kicking phase.

The positive square root of the adjusted coefficient of multiple determination, which is the adjusted coefficient of multiple correlation (CMC, 0 < CMC < 1) [21,22], was calculated for evaluating the similarity of force–time curves. One comparison was between F_{net} and F force–time curves. The similarity between F_{net} and F was represented by CMC_{net}. The mean force difference and mean absolute force difference between F_{net} and F were $M_{F_{net}-F}$ and $M_{|F_{net}-F|}$. Another comparison was between F_{pro} and F force–time curves. The similarity between F_{pro} and F was represented by CMC_{pro}. The mean force difference and mean absolute force difference between F_{pro} and F were $M_{F_{pro}-F}$ and $M_{|F_{pro}-F|}$. The mean force differences and mean absolute force differences were computed over force curves averaged over 10 force-producing phases. The mean force differences, which are $M_{F_{net}-F}$ and $M_{F_{pro}-F}$, were calculated to provide descriptive statistics only. The forces in this study were presented as values relative to body weight (%BW).

2.5. Statistical Analyses

A two-way mixed factorial ANOVA was performed. The dependent variables of the analyses were (1) CMCs and (2) mean absolute force differences. The independent variables were the speed (or the incline) of the treadmill and the comparisons (i.e., comparison between F_{net} and F and comparison between F_{pro} and F). The speed (or the incline) of the treadmill was treated as the within-subject factor, and the comparison pair was treated as the between-subject factor. The EMMEANS subcommand with the Bonferroni adjustment in SPSS was used to perform the pairwise comparisons of the dependent variable when interactions were detected [23], and the effect size ($_p\eta^2$) was calculated for further evaluation. The level of statistical significance was set at $\alpha = 0.05$. All data are presented as mean $\pm$ standard deviation (SD). Data analyses were conducted using version 23.0 of the SPSS program package for statistical analysis (SPSS Inc., Chicago, IL, USA).

3. Results

The force–time curves of F, F_{net}, and F_{pro} for the DP and the V2 techniques are shown in Figure 6. The interaction effect (comparison * speed) was significant on CMC with the DP technique ($p = 0.038$), but not with the V2 technique ($p = 0.988$). CMC_{pro} did not differ from CMC_{net} at any speed in the DP technique ($p \geq 0.106$, Table 1). With the V2 technique, the overall CMC_{pro} was about 5% lower than CMC_{net} ($p = 0.011$, Table 1). The interaction effect (comparison * incline) was not significant on CMC in the DP technique ($p = 0.620$) but was significant in the V2 technique ($p = 0.042$). In the DP technique, the main effect of comparison on CMC was not significant ($p = 0.218$, Table 1). In the V2 technique, CMC_{net} was significantly greater than CMC_{pro} at 3° ($p = 0.042$, Table 1).

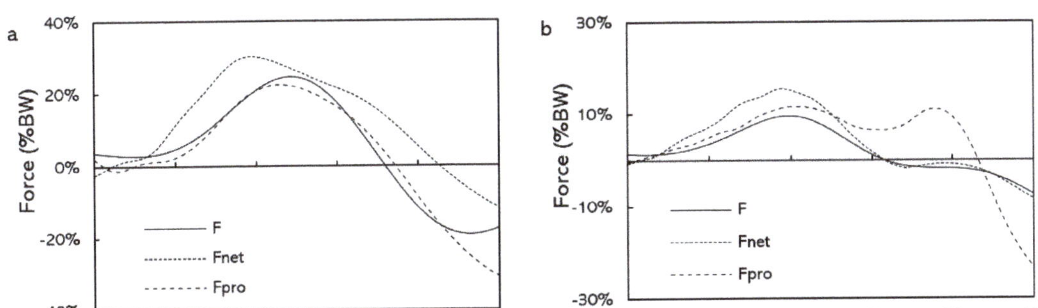

Figure 6. Force-time curves of F, F_{net}, and F_{pro}: (**a**) DP technique, (**b**) V2 technique. Values are averaged over 10 force-producing phases of one subject from each technique (speed of the treadmill was 19 km/h; incline of the treadmill was 2°).

Table 1. Mean and standard deviation of the CMC for the DP technique (n = 9) and the V2 technique (n = 10).

		DP Technique				V2 Technique			
		CMC_{net}	CMC_{pro}	p-Value	$P\eta^2$	CMC_{net}	CMC_{pro}	p-Value	$P\eta^2$
Speeds	13 km/h	0.935 ± 0.022	0.910 ± 0.038	0.106 [b]	0.155	0.901 ± 0.048	0.853 ± 0.043		
	15 km/h	0.933 ± 0.023	0.916 ± 0.034	0.230 [b]	0.089	0.908 ± 0.047	0.862 ± 0.050		
	17 km/h	0.920 ± 0.030	0.919 ± 0.030	0.951 [b]	0.001	0.905 ± 0.040	0.861 ± 0.035	0.011 [a]	0.309
	19 km/h	0.901 ± 0.045	0.908 ± 0.046	0.778 [b]	0.005	0.885 ± 0.045	0.837 ± 0.047		
	21 km/h	0.883 ± 0.058 [1,2,3]	0.907 ± 0.042	0.330 [b]	0.059	0.879 ± 0.044	0.832 ± 0.041		
	p-value	0.043 [d]	0.371 [d]			0.008 [c]			
	$P\eta^2$	0.509	0.264			0.216			
Inclines	3°	0.933 ± 0.024	0.914 ± 0.046			0.911 ± 0.032	0.856 ± 0.073	0.042 [b]	0.210
	4°	0.946 ± 0.016	0.932 ± 0.033	0.218 [a]	0.093	0.922 ± 0.041	0.896 ± 0.044 *	0.179 [b]	0.098
	5°	0.955 ± 0.015	0.936 ± 0.037			0.912 ± 0.047	0.900 ± 0.055 *	0.617 [b]	0.014
	p-value	0.001 [e]				0.479 [f]	0.007 [f]		
	$P\eta^2$	0.464				0.083	0.446		

Note: CMC_{net} represents the similarity between F and F_{net}. CMC_{pro} represents the similarity between F and F_{pro}. [a] p-value for the main effect of comparison in a two-way mixed factorial ANOVA. [b] p-value for pairwise comparisons when interactions were detected. [c] p-value for the main effect of speed in a two-way mixed factorial ANOVA. [d] p-value for the simple effect of speed when interactions were detected. [e] p-value for the main effect of incline in a two-way mixed factorial ANOVA. [f] p-value for the simple effect of incline when interactions were detected. [1] Significantly different from 13 km/h. [2] Significantly different from 15 km/h. [3] Significantly different from 17 km/h. * Significantly different from 3°.

On average, the $M_{F_{net}-F}$ was lower than zero and the $M_{F_{pro}-F}$ was greater than zero (Figure 7) for both the DP and V2 techniques at any speeds and any inclines. The interaction effect (comparison * speed) was significant on the absolute mean force difference with the DP technique (p = 0.025) but not with the V2 technique (p = 0.165). In the DP technique, $M_{|F_{net}-F|}$ was 24% lower than $M_{|F_{pro}-F|}$ at 15 km/h (p = 0.028, Table 2). For the V2 technique, the overall $M_{|F_{pro}-F|}$ was about 37% greater than $M_{|F_{net}-F|}$. The interaction effect (comparison * incline) was not significant on absolute mean force difference in the DP technique (p = 0.393) but was significant in the V2 technique (p = 0.016). In the DP technique, the overall $M_{|F_{net}-F|}$ was about 39% lower than $M_{|F_{pro}-F|}$. With the V2 technique, $M_{|F_{net}-F|}$ was significantly lower than $M_{|F_{pro}-F|}$ at 3° and 4° ($p \leq 0.013$, Table 2).

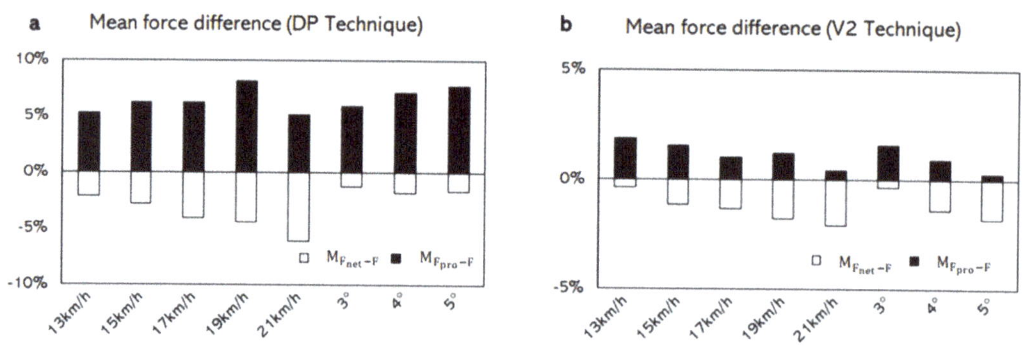

Figure 7. Mean force difference over force producing phases in DP and V2 techniques (%BW). $M_{F_{net}-F}$ represents the difference between F and F_{net} and is calculated by F–F_{net}. $M_{F_{net}-F}$ lower than zero indicates that F_{net} is greater than F. $M_{F_{pro}-F}$ represents the difference between F and F_{pro} and is calculated by F–F_{pro}. $M_{F_{pro}-F}$ greater than zero indicates that F_{pro} is lower than F.

Table 2. Mean and standard deviation of the mean absolute difference for the DP technique (n = 9) and V2 technique (n = 10) (BW%).

		DP Technique				V2 Technique			
		$M_{\lvert F_{net}-F \rvert}$	$M_{\lvert F_{pro}-F \rvert}$	p-Value	$P\eta^2$	$M_{\lvert F_{net}-F \rvert}$	$M_{\lvert F_{pro}-F \rvert}$	p-Value	$P\eta^2$
Speed	13 km/h	6.1 ± 1.1	8.1 ± 2.9	0.058 [b]	0.207	2.9 ± 0.4	4.4 ± 0.8		
	15 km/h	6.9 ± 1.1	9.1 ± 2.6 [4]	0.028 [b]	0.268	3.1 ± 0.5	4.7 ± 0.4		
	17 km/h	8.5 ± 1.5 [1,2,5]	10.2 ± 3.3 [1,4]	0.166 [b]	0.116	3.6 ± 0.6	4.8 ± 0.5	0.001 [a]	0.633
	19 km/h	9.0 ± 1.3 [1,2]	11.6 ± 3.6 [1,2,3]	0.057 [b]	0.209	4.0 ± 0.9	5.4 ± 0.8		
	21 km/h	10.8 ± 2.2 [1,2,3]	10.9 ± 2.4 [1]	0.992 [b]	0.001	4.4 ± 0.9	5.3 ± 0.5		
	p-value	0.001 [d]	0.001 [d]			0.001 [c]			
	$P\eta^2$	0.856	0.857			0.588			
Inclines	3°	6.2 ± 0.9	8.7 ± 3.0			2.9 ± 0.5	4.5 ± 0.8	0.001 [b]	0.617
	4°	7.1 ± 1.1	9.6 ± 2.7	0.015 [a]	0.315	3.4 ± 1.1	4.5 ± 0.6	0.013 [b]	0.295
	5°	7.6 ± 1.3	10.7 ± 2.9			3.7 ± 0.7 *	4.3 ± 0.9	0.115 [b]	0.132
	p-value	0.001 [e]				0.014 [f]	0.577 [f]		
	$P\eta^2$	0.615				0.394	0.063		

Note: $M_{\lvert F_{net}-F \rvert}$ represents the absolute difference between F and F_{net} and is calculated by $\lvert F - F_{net} \rvert$. $M_{\lvert F_{pro}-F \rvert}$ represents the absolute difference between F and F_{pro} and is calculated by $\lvert F - F_{pro} \rvert$. [a] p-value for the main effect of comparison in a two-way mixed factorial ANOVA. [b] p-value for pairwise comparisons when interactions were detected. [c] p-value for the main effect of speed in a two-way mixed factorial ANOVA. [d] p-value for the simple effect of speed when interactions were detected. [e] p-value for the main effect of incline in a two-way mixed factorial ANOVA. [f] p-value for the simple effect of incline when interactions were detected. [1] Significantly different from 13 km/h. [2] Significantly different from 15 km/h. [3] Significantly different from 17 km/h. [4] Significantly different from 19 km/h. [5] Significantly different from 21 km/h. * Significantly different from 3°.

With the DP technique, CMC_{pro} was independent from the speed (p = 0.371, Table 1). However, CMC_{net} decreased significantly at 21 km/h when compared to CMC_{net} at 13, 15, and 17 km/h ($p \leq 0.046$). The overall CMC increased by about 2% from 3 to 5°. Both $M_{\lvert F_{net}-F \rvert}$ and $M_{\lvert F_{pro}-F \rvert}$ increased with the increasing speed of the treadmill ($p < 0.001$, $p < 0.001$, Table 2). The overall absolute mean difference increased by 23% from 3 to 5°. With the V2 technique, the overall CMC decreased by about 2% from 13 to 21 km/h. CMC_{net} was independent of the incline of the treadmill (p = 0.042, Table 1). CMC_{pro} increased from 3 to 5° (p = 0.007, Table 1). The overall absolute difference increased by 33% from 13 to 21 km/h. $M_{\lvert F_{net}-F \rvert}$ was dependent on the incline of the treadmill (p = 0.014, Table 2), and $M_{\lvert F_{pro}-F \rvert}$ was independent of the incline of the treadmill (p = 0.577, Table 2).

4. Discussion

The results of this study support our first hypothesis that the force-time curves of F_{net} and F_{pro} all give comparable shape with F in both techniques. In the DP technique, CMC_{pro} ranged from 0.907 to 0.936, CMC_{net} ranged from 0.883 to 0.955 and did not differ from CMC_{pro} (Table 1). In the V2 technique, CMC_{pro} ranged from 0.832 to 0.900, and CMC_{net} ranged from 0.879 to 0.922 (Table 1). The CMC depicting the similarity between waveforms and CMC close to 1 indicated that the curves involved were similar [21,22]. Therefore, the shapes of force-time curves of F_{pro} and F_{net} all showed similar to force-time curves of F, and both could be used to describe the shape of F during the poling phase of the DP technique and the kicking phase of the V2 technique. In addition, in the V2 technique, CMC_{net} was 5% higher than CMC_{pro} while changing the speed (Table 1), indicating that the force-time curves of F_{net} was more comparable to the force-time curves of F than F_{pro} while using the V2 technique. Consequently, F_{net} appears to be more appropriate for determining the trend of the forward acceleration in the V2 technique.

The results of this study partly support our second hypothesis that F_{net} would give a considerable overestimation and F_{pro} would be more accurate than F_{net} when estimating F in both the DP and V2 techniques. In this present study, the F_{net} had a considerable overestimation when estimating F in both the DP and V2 techniques, but F_{pro} was not more accurate than F_{net} in both techniques. The mean force differences over force curves between F_{pro} and F ($M_{F_{pro}-F}$), as well as F_{net} and F ($M_{F_{net}-F}$), were computed (Figure 7).

The mean force differences in this study indicated that, on average, F_{net} would overestimate ($M_{F_{net}-F} < 0$, Figure 7) the F in both the DP and V2 techniques. F_{net} was calculated from the GRF directly. The costs associated with the transformations of energy [24] between each segment and the elastic potential energy of the muscle were not taken out from F_{net}. Thus, a considerable difference in F_{net} and F may exist. The F_{pro} underestimate ($M_{F_{pro}-F} > 0$, Figure 7) the F, but it was not more accurate than F_{net} in both techniques. F_{pro} was calculated by combining the GRF and the position of COM. The resultant GRF was subdivided into a translational component, which acted through the COM, and a rotational component, which was always perpendicular to the translational component [6,25]. Because the rotational component will not have a translational effect on the COM, when F_{pro} was calculated, the rotational component was not involved. Therefore, the forward component of the translational component might underestimate the forward acceleration in both the DP and V2 techniques. The absolute mean force differences (Table 2) were computed to evaluate which force-time curve was closer to the reference one. A smaller absolute mean force difference indicates a force-time curve closer to the reference curve and further shows a relatively higher accuracy. The results of this study showed that with both the DP and V2 techniques, the absolute mean force difference between F_{pro} and F were greater than or have no difference with the absolute mean force difference between F_{net} and F. This indicates that the force-time curves of F_{net} were closer to or have no difference with the force–time curves of F. Thus, F_{pro} was not more accurate than F_{net}.

The results of this study do not support our third hypothesis that the approaches to calculate the F_{net} and F_{pro} would not be affected by the speed and incline of the treadmill in both techniques. The approaches to calculate the F_{net} and F_{pro} were all influenced by the speed or the incline of the treadmill. As there was a balance of forces under laboratory conditions with no air resistance, constant friction coefficient, and constant gravitational force, the total external force remained constant when the speed was changed. The gravity component parallel to the treadmill surface increased with the incline [19]; thus, more forces were needed at a steeper incline. It is impossible to have an exact reproduction of the reference value F; however, if the methods for calculating F_{net} and F_{pro} were independent from the speed and the incline of the treadmill, the CMC_{net}, CMC_{pro}, and the absolute mean force difference over the force-generating cycle should remain constant in both techniques. The CMCs in this study were somehow affected by the speed and incline of the treadmill in both the DP and V2 techniques. In addition, the results of this study showed that the absolute mean force differences between F_{net} and F ($M_{|F_{net}-F|}$) and between F_{pro} and F ($M_{|F_{pro}-F|}$) were all affected by the speed of the treadmill regardless of whether the DP or V2 technique was performed (Table 2). The absolute mean force differences increased with increasing speed (Table 2), which means that although F_{pro} and F_{net} can be used to estimate the force-time curve of F, they do not remain stable when the speed changes. Thus, when investigating how F adapts to increasing speed by using F_{net} or F_{pro}, the increasing mean force differences should be considered. Both $M_{|F_{net}-F|}$ and $M_{|F_{pro}-F|}$ increased when the DP technique was used while increasing the incline of the treadmill (Table 2). However, when the V2 technique was used, $M_{|F_{net}-F|}$ was affected by the increasing incline, and the significant increase was only found at the steepest incline (Table 2), but $M_{|F_{pro}-F|}$ was not influenced by the incline of the treadmill. Thus, compared to F_{net}, F_{pro} was more stable when estimating F while changing the incline of the treadmill.

Therefore, when considering the whole poling phase in the DP technique, both F_{pro} and F_{net} are appropriate for estimating the trend of F. The similarity between the F_{pro} and F is stable while changing the speed in the DP technique. However, F_{net} has better accuracy than F_{pro} when the speed and the incline is changed. When considering the whole kicking phase in the V2 technique, the trend of F_{net} fits F better. However, the similarity between the F_{pro} and F is stable in the V2 technique when the incline is changed. As the result in the DP technique, F_{net} also has better accuracy than F_{pro} in the V2 technique. There are some limitations of this study. The calculation of the COM is dependent on the assumed mass distributions. Although this has been proved to be suitable for estimating the position of

COM in sports, it can still cause the golden standard of the reference to be inaccurate. In addition, the PFAs of the leg force and pole force were estimated points, and this may also have some effects on the accuracy of F_{pro}. Furthermore, the added measurement equipment could have affected skiing performance.

5. Conclusions

The present study evaluated two approaches for estimating the total propulsive force on skier's COM. Both approaches can estimate the trend of the force-time curve of the propulsive force properly. Although both had a considerable overestimation; an approach by calculating the forward-directed horizontal component of 3D GRF is a more appropriate method due to a better accuracy. Future studies could investigate the contribution of skis and poles to forward COM acceleration by calculating the propulsive force from skis and poles separately. Moreover, as for the gliding phase that exists in XC skiing, the velocity at the end of the force generating phase is important. Future studies could also investigate the contributions of skis and poles to velocity change separately.

Author Contributions: Conceptualization, S.Z., O.O., K.R., L.K., S.L., C.G. and V.L.; Data curation, S.Z.; Formal analysis, S.Z.; Funding acquisition, S.Z. and V.L.; Investigation, S.Z., O.O., K.R., L.K., S.L., C.G. and V.L.; Methodology, S.Z., O.O., K.R., L.K., S.L., C.G. and V.L.; Project administration, S.Z.; Resources, S.Z., O.O. and K.R.; Supervision, S.Z., O.O., S.L. and V.L.; Validation, S.Z.; Visualization, S.Z.; Writing—original draft, S.Z.; Writing—review and editing, S.Z., O.O., K.R., L.K., S.L., C.G. and V.L. All authors have read and agreed to the published version of the manuscript.

Funding: This research was funded by the China Scholarship Council, grant number 201806520003.

Institutional Review Board Statement: The study was conducted according to the guidelines of the Declaration of Helsinki and approved by the Ethics Committee of the University of Jyväskylä.

Informed Consent Statement: Informed consent was obtained from all subjects involved in the study. Written informed consent was obtained from the patient(s) to publish this paper.

Data Availability Statement: The data presented in this study are available on request from the corresponding author. The data are not publicly available because the data also form part of ongoing studies.

Acknowledgments: The author Shuang Zhao wants to thank Christina Mishica and Ritva Mikkonen for language optimization. The authors would like to express their appreciation and thanks to the staff, athletes, and coaches for their participation, enthusiasm, and cooperation in this study.

Conflicts of Interest: The authors declare no conflict of interest.

Appendix A. Measurement and Calculation of Resistance Friction Coefficient of Roller Ski

The resistance friction coefficient of roller ski was measured on the treadmill surface using a custom-made friction measurement device (University of Jyväskylä, Jyväskylä, Finland, Figure A1) and calculated with the LabVIEW software package (National Instruments, Austin, TX, USA) before the measurement. A commercial force sensor (Raute precision TB5, Nastola, Finland) that measures the anterior–posterior force along the roller ski (F_Y) was contained in the friction measurement device. The friction coefficient between the treadmill surface and the roller ski was obtained by $\mu = \frac{F_Y}{F_Z}$, where F_Z is the vertical force that equals the weight of the weight plate placed on the roller ski.

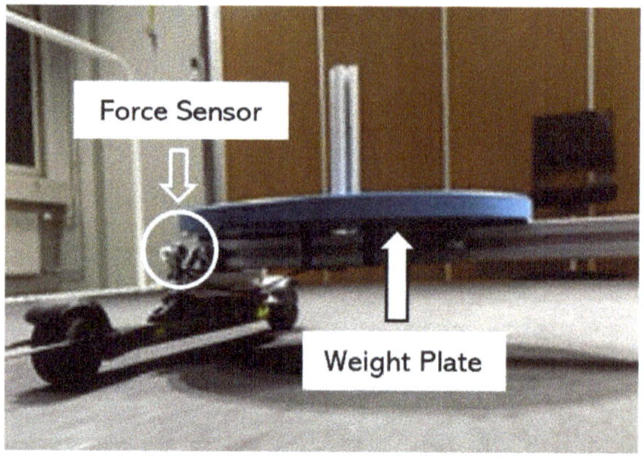

Figure A1. Custom-made friction measurement device.

Appendix B. Mechanical Principle of Translational Force

The motion of a rigid body under external forces can be reduced to (i) the acceleration of the COM and to (ii) the angular acceleration of the object around its COM. These change the rate of momentum and angular momentum, respectively. Correspondingly, in a 3D space, the motion problem involves six degrees of freedom (DoF). These DoFs can be expressed with three components of a translational force and another three components of a moment. The forces and torques make a rigid body translate and rotate. It is worth noting that it is a modeler's decision to express the six DoFs with translational forces and torques with respect to the COM. Accordingly, this is not the only, but rather a practical, option to model motion. The decomposition of an external force into translational force and torque components acting on a rigid object is illustrated in Figure B1. An external force $\overrightarrow{F_{resultant}}$ acts on point a of a rigid sphere. $\overrightarrow{F_{resultant}}$ is decomposed to the translational component $\overrightarrow{F_{translational}}$ that acts in the direction of the line joining the COM and point a. The (displacement) vector from COM to point a is denoted by l. The rotational component $\overrightarrow{F_{rotational}}$ of the force is perpendicular to vector l such that condition $\overrightarrow{F_{resultant}} = \overrightarrow{F_{rotational}} + \overrightarrow{F_{translational}}$ holds (Figure B1a). Precisely, the same situation is expressed in terms of a translational force $\overrightarrow{F_{translational}}$ and torque τ, which is the product of l and $\overrightarrow{F_{rotational}}$ (Figure B1b).

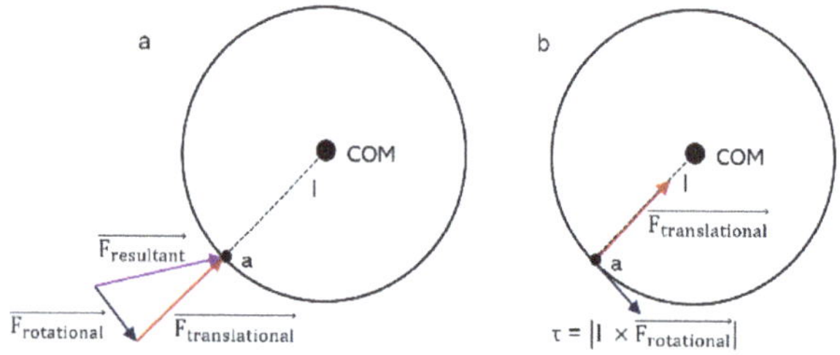

Figure B1. Diagram of the mechanical principle of translational force.

As the principles of mechanics do not depend on the object—that is, Newton's laws apply to all objects—the basic setting does not change from that of a single rigid object. However, in the case of joined bodies, the COM cannot be specified a priori, as it depends on the position of the parts in relation to each other (see Figure B2). When external forces

$\overrightarrow{F_{resultant}}$ act on the object (Figure B2a), the whole object translates, and each part moves with respect to each other. Because of this, the COM moves with respect to the parts. However, the motion of the object still fulfils Newton's laws (Figure B2b). Consequently, nothing prevents one from decomposing the external forces into components of translational forces and torques with respect to the COM.

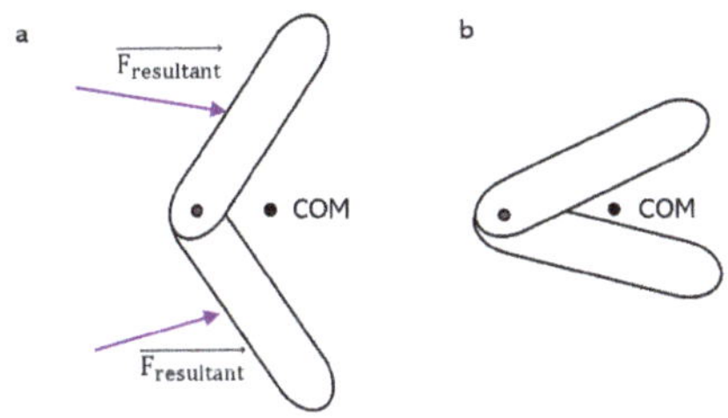

Figure B2. Diagram of forces acting on connected rigid bodies.

References

1. Smith, G.A. Biomechanics of cross country skiing. In *Cross Country Skiing. Handbook of Sports Medicine*; Rusko, H., Ed.; Wiley: New York, NY, USA, 2003; pp. 32–61.
2. Mapelli, A.; Zago, M.; Fusini, L.; Galante, D.; Colombo, A.; Sforza, C. Validation of a protocol for the estimation of three-dimensional body center of mass kinematics in sport. *Gait Posture* **2014**, *39*, 460–465. [CrossRef] [PubMed]
3. Stoggl, T.; Holmberg, H.C. Three-dimensional Force and Kinematic Interactions in V1 Skating at High Speeds. *Med. Sci. Sports Exerc.* **2015**, *47*, 1232–1242. [CrossRef] [PubMed]
4. Smith, G.; Kvamme, B.; Jakobsen, V. Ski skating technique choice: Mechanical and physiological factors affecting performance. In Proceedings of the ISBS-Conference Proceedings Archive, Ouro Preto, Brazil, 23–27 August 2007.
5. Hoset, M.; Rognstad, A.; Rølvåg, T.; Ettema, G.; Sandbakk, Ø. Construction of an instrumented roller ski and validation of three-dimensional forces in the skating technique. *Sports Eng.* **2014**, *17*, 23–32. [CrossRef]
6. Göpfert, C.; Pohjola, M.V.; Linnamo, V.; Ohtonen, O.; Rapp, W.; Lindinger, S.J. Forward acceleration of the centre of mass during ski skating calculated from force and motion capture data. *Sports Eng.* **2017**, *20*, 141–153. [CrossRef]
7. Ohtonen, O.; Lindinger, S.J.; Gopfert, C.; Rapp, W.; Linnamo, V. Changes in biomechanics of skiing at maximal velocity caused by simulated 20-km skiing race using V2 skating technique. *Scand. J. Med. Sci. Sports* **2018**, *28*, 479–486. [CrossRef] [PubMed]
8. Smith, G.A. Biomechanical analysis of cross-country skiing techniques. *Med. Sci. Sports Exerc.* **1992**, *24*, 1015–1022. [CrossRef] [PubMed]
9. Holmberg, H.C.; Lindinger, S.; Stöggl, T.; Björklund, G.; Müller, E. Contribution of the legs to double-poling performance in elite cross-country skiers. *Med. Sci. Sports Exerc.* **2006**, *38*, 1853–1860. [CrossRef] [PubMed]
10. Andersson, E.; Stoggl, T.; Pellegrini, B.; Sandbakk, O.; Ettema, G.; Holmberg, H.C. Biomechanical analysis of the herringbone technique as employed by elite cross-country skiers. *Scand. J. Med. Sci. Sports* **2014**, *24*, 542–552. [CrossRef] [PubMed]
11. Stoggl, T.; Muller, E.; Lindinger, S. Biomechanical comparison of the double-push technique and the conventional skate skiing technique in cross-country sprint skiing. *J. Sports Sci.* **2008**, *26*, 1225–1233. [CrossRef] [PubMed]
12. Stoggl, T.L.; Holmberg, H.C. Double-Poling Biomechanics of Elite Cross-country Skiers: Flat versus Uphill Terrain. *Med. Sci. Sports Exerc.* **2016**, *48*, 1580–1589. [CrossRef] [PubMed]
13. Sandbakk, O.; Holmberg, H.C. A reappraisal of success factors for Olympic cross-country skiing. *Int. J. Sports Physiol. Perform.* **2014**, *9*, 117–121. [CrossRef] [PubMed]
14. Stöggl, T.; Stöggl, J.; Müller, E. Competition analysis of the last decade (1996–2008) in crosscountry skiing. In *Science and Skiing IV*; Erich Müller, S.L., Thomas, S., Eds.; Meyer & Meyer Sport: Berkshire, UK, 2008; pp. 657–677.
15. Ohtonen, O.; Lindinger, S.; Lemmettylä, T.; Seppälä, S.; Linnamo, V. Validation of portable 2D force binding systems for cross-country skiing. *Sports Eng.* **2013**, *16*, 281–296. [CrossRef]
16. Ohtonen, O.; Ruotsalainen, K.; Mikkonen, P.; Heikkinen, T.; Hakkarainen, A.; Leppävuori, A.; Linnamo, V. Online feedback system for athletes and coaches. In Proceedings of the 3rd International Congress on Science and Nordic Skiing, Vuokatti, Finland, 5–8 June 2015; p. 35.
17. Yu, B.; Gabriel, D.; Noble, L.; An, K.-N. Estimate of the optimum cutoff frequency for the Butterworth low-pass digital filter. *J. Appl. Biomech.* **1999**, *15*, 318–329. [CrossRef]
18. Selbie, W.S.; Hamill, J.; Kepple, M.T. Three-Dimentional Kinetics. In *Research Methods in Biomechanics*, 2nd ed.; Robertson, G.E., Caldwell, G.E., Hamill, J., Kamen, G., Whittlesey, S., Eds.; Human Kinetics: Champaign, IL, USA, 2013; pp. 159–160.
19. Danielsen, J.; Sandbakk, Ø.; McGhie, D.; Ettema, G. Mechanical energetics and dynamics of uphill double-poling on roller-skis at different incline-speed combinations. *PLoS ONE* **2019**, *14*, e0212500. [CrossRef] [PubMed]
20. Winter, D.A. Human balance and posture control during standing and walking. *Gait Posture* **1995**, *3*, 193–214. [CrossRef]

21. Kadaba, M.; Ramakrishnan, H.; Wootten, M.; Gainey, J.; Gorton, G.; Cochran, G. Repeatability of kinematic, kinetic, and electromyographic data in normal adult gait. *J. Orthop. Res.* **1989**, *7*, 849–860. [CrossRef] [PubMed]
22. Yu, B.; Kienbacher, T.; Growney, E.S.; Johnson, M.E.; An, K.N. Reproducibility of the kinematics and kinetics of the lower extremity during normal stair-climbing. *J. Orthop. Res.* **1997**, *15*, 348–352. [CrossRef] [PubMed]
23. Malek, M.H.; Coburn, J.W.; Marelich, W.D. *Advanced Statistics for Kinesiology and Exercise Science: A Practical Guide to ANOVA and Regression Analyses*; Routledge: Oxfordshire, UK, 2018.
24. Robertson, G.E. Engergy, Work, and Power. In *Research Methods in Biomechanics*, 2nd ed.; Robertson, G.E., Caldwell, G.E., Hamill, J., Kamen, G., Whittlesey, S., Eds.; Human Kinetics: Champaign, IL, USA, 2013; p. 132.
25. Schwameder, H. Biomechanics research in ski jumping, 1991–2006. *Sports Biomech.* **2008**, *7*, 114–136. [CrossRef] [PubMed]

sensors

MDPI

Article

A Wearable System for Jump Detection in Inline Figure Skating

Antonio Panfili *, **Alvise Spanò** * **and Agostino Cortesi** *

Dipartimento di Scienze Ambientali, Informatica e Statistica, Università Ca' Foscari Venezia, 30172 Venice, Italy
* Correspondence: antoniopanfili@gmail.com (A.P.); alvise.spano@unive.it (A.S.); cortesi@unive.it (A.C.)

Abstract: This article presents the design and experimental evaluation of a non-invasive wearable sensor system that can be used to acquire crucial information about athletes' performance during inline figure skating training. By combining distance and time-of-flight sensors and gyroscopes, the system is able to detect when jumps are performed and provides a live view of the data (e.g., the number and height of jumps) through a graphical user interface. The main novelty of our approach lies in the way in which the optical sensors are orientated. Typically, the sensors are orientated horizontally and positioned in pairs on the ground, where they measure the time interval between the moment the athlete leaves the ground and the moment they land. In our system, an optical sensor is placed under each foot and is vertically orientated so as to constantly measure the distance from the ground. In addition, a gyroscope sensor is placed on the athlete's back, which provides information on the direction and angular momentum of the movement. By combining this data, the system provides the accurate detection of various jumps and technical elements without any constraints on the training ground. In this paper, the system is also compared to similar platforms in the literature, although there are no other specific systems that are available for inline figure skating. The results of the experimental evaluation, which was performed by high profile athletes, confirm its effectiveness in correctly detecting jumps, especially considering its compromise between precision and the overall cost of the equipment.

Keywords: wearable sensors; inline figure skating; sports biomechanics; human movement analysis

Citation: Panfili, A.; Spanò, A.; Cortesi, A. A Wearable System for Jump Detection in Inline Figure Skating. *Sensors* **2022**, *22*, 1650. https://doi.org/10.3390/s22041650

Academic Editor: Vesa Linnamo

Received: 20 December 2021
Accepted: 16 February 2022
Published: 20 February 2022

1. Introduction

The availability of more precise sensors and more efficient data transmission technologies has allowed the development of systems that can be used to support the specialised technical preparation of athletes for different sports.

This work turns its attention to the sport of inline figure skating, for which sensor-based systems that are as sophisticated as those used in other sports disciplines have not yet been developed. Inline figure skaters train using roller skates, where the wheels are arranged on a single line and a piece of rubber at the front emulates the tip of the figure skating blade. In the future, our system could also be adapted for use in figure skating on ice and other winter sports as well [1].

1.1. Scenario

Firstly, we will provide some basic notions concerning equipment and athletic gestures, which we refer to as technical elements, or simply elements, from now on. To introduce the basic concepts relating to the technical elements, reference is made to a technical regulation, namely the ISU (International Skating Union) for the 2019/2020 competitive season [2] (for a comprehensive list of acronyms, see the table in the Appendix A). This choice was based on the quality and precision of how the elements are defined, as well as on the historical background of the ISU regulation that was approved during the 50th annual ISU congress in Scheveningen in the Netherlands, which is still in use today with minor changes carried out annually. Furthermore, the regulation adopted by the FIRS (Fédération International de Roller Sports), known as WS (World Skate) since the 2016/17 and 2017/18 competitive seasons, for inline figure skating is very similar to the ISU regulation. The final score of a competition programme is divided into the *technical element score* and the *programme component score*. The types of technical elements are: jumps, spins, step sequences and choreographic sequences. Figure skating jumps are grouped into different *families*: Salchow, toe loop, loop, flip, Lutz and Axel. Exceptions include: the *waltz* jump is only taken into account for youth categories because of its reduced rotation (half turn); the *Euler* jump (abbreviation: 1Eu), which is similar to a loop in that the landing foot has to be the opposite of the starting foot but is only considered when it is executed within a combination of jumps, i.e., between two jumps of the *families* listed above. The Euler, when used in combination between two listed jumps, becomes a listed jump (1Eu) with its value indicated in the SOV (scale of values) in the Technical Panel Handbook [3].

Notably, in the proposed sensor system, the type of jump is not taken into consideration; only the number of jumps and their height count. Each jump, excluding the *waltz*, must be performed with 1–4 rotations. Another exception is the *Axel*, for which jumps must be performed with $1\frac{1}{2}$–$4\frac{1}{2}$ rotations [3].

1.2. Challenge

The main challenge was to design a non-invasive wearable sensor system that could capture crucial information about an athlete's performance during training. This information then needs to be processed to monitor and improve the relevant parameters of the training session. In particular, the following constraints were taken into consideration:

- the wearable device that is equipped with the sensors must not hinder the athlete or compromise their freedom of movement in any way;
- the system must not restrict the athletes to a limited area or to a specific surface with particular materials or lighting conditions;
- the number of athletes training simultaneously must not jeopardise the effectiveness of the system;
- the overall accuracy of the system determines the reliability of the product, hence the error correction applied to the raw data from the sensors is crucial;
- the cost is a relevant and decisive aspect for the choice of hardware of which the system is composed.

The detection of the technical elements is probably the most difficult task. It is not enough to detect whether an athlete has come off the ground and then measure the vertical distance: the real challenge is to filter out any vertical movements that would not be considered a valid jump or a technical element in a real performance.

1.3. Objectives

Our goal was to create a wearable device that is capable of accurately detecting, measuring and counting jumps performed by athletes when training for inline figure skating. The intended audience are, therefore, athletes during training sessions and not official competitions. We show that a system of non-invasive sensors can improve the effectiveness of training sessions by collecting data that are then analysed offline to provide further insights into performance.

As a minor objective, this information should be easy to access through a user-friendly graphic interface.

1.4. Contribution

The world of inline figure skating lacks the significant technological support that aims, above all, to provide athletes with useful information to improve their training sessions. Therefore, any automated feedback on the most relevant athletic elements is a valuable contribution to this field. From a scientific point of view, the main contribution of this study consists of carrying out the accurate measurement and detection of various jumps and technical elements by means of a single, non-invasive and inexpensive device that is equipped with sensors. No intervention to the training ground or environment is necessary.

The main novelty of this work is in the way that the optical sensors are orientated. Typically, sensors are orientated horizontally and placed in pairs on the ground to measure the time interval between the moment the athlete takes off from the ground and the moment they land again, as with those listed in Section 5.3. Our approach instead consists of installing an optical sensor under each foot and orientating it vertically in order to measure the foot's distance from the ground constantly by sampling it several times per second. Additionally, one gyroscope–accelerometer pair is installed on the back of the athlete, providing a further source of information: namely, the direction and the angular momentum of motion. All data can then be combined and processed in real time to detect and count the jumps and to calculate the height and other parameters of the parabolic curve performed by the athlete.

1.5. System Architecture

The overall architecture of our proposed system is depicted in Figure 1. It involves the use of distance and time-of-flight sensors and a gyroscope connected to a small computer that has the task of retrieving the sensors' data and storing them for subsequent evaluation through a custom application, which is also discussed in this paper.

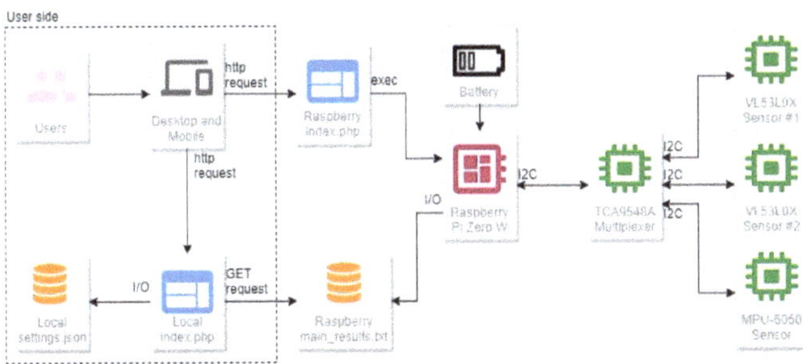

Figure 1. An overview of the system's software and hardware components.

The system consists of two laser sensors placed under the skates (inside the frame) and a control unit placed on the back of the athlete, composed of an RPiZW board and a triaxial gyroscope. The two laser sensors and the gyroscope are connected to the control unit via cables passing through a multiplexer and the control unit is powered by a battery. When both the control unit and the user's device are connected to the same network, it is possible for the user to start the sampling process using an easy and responsive GUI. When the sampling process is run, the control unit starts recording all sensors' data into a shared text file in CSV format. At that point, the user can view a web page that automatically obtains all of the data from the shared file and, after the analysis, can show them all of the data either in real time (the updates run every 3 s) or in historic mode. Through this analysis web page, the user can edit some of the settings, such as the chart height, and add the IP address of the control unit.

2. Materials and Methods

In this section, we provide details of how we conducted our study and designed our system. In inline figure skating, one of the crucial factors is to optimise the jumps in any possible way, as they are the basic constituents of the technical elements. The ability to analyse the jumps in detail is a priority of our proposal: as shown in Figure 1, the two optical sensors installed under each skate produce enough data to draw two graphs, one for each foot, which show the athlete's distance from the ground as a function of time, sampled about 10 times per second.

One important aspect of our system is that it does not attempt to infer the family of a jump—e.g., Lutz, Axel, Euler, etc. as listed in Section 1.1—but it detects any jump in the general sense by sampling the athlete's distance from the ground. A jump is only detected when a given threshold is crossed several times in a row within a limited time interval. The algorithm and the detection strategy will be detailed in Sections 2.4 and 3. Note that identifying the family of a jump was deemed as an unnecessary complication that, in the end, would not be worth the effort, as the athlete's physical and biological parameters need to be taken into account to obtain precise results, such as their weight, height, dominant hand/foot, etc.

Definition 1 (Detachment). *We considered the athlete as detached from the ground when the distance measured by both of the laser sensors between the skate and the ground surpassed a given threshold.*

This *threshold* was carefully calibrated through extensive testing and set by default to 20 cm. We calibrated this value by considering the distance between the sensor and the ground during the phase of the maximum inclination of the foot, including the front inclination, which was performed by climbing on both brakes by raising all three wheels, and the lateral inclination, which occurred during the execution of deep edges. This threshold did not depend on the athlete's height nor on the size of their foot since the height of the frame was the same regardless of its length. However, this value could be modified through the user interface of the program in order to support changes in frame size, changes in wheel diameter, etc.

Definition 2 (Jump). *We defined a jump as the contiguous series of N distance measurements performed by the laser sensors when they were detached from the ground, according to Definition 1.*

In this study, the value of $N = 3$ was hard-coded into our algorithm as it yielded the best results according to our testing.

Our system was a compromise between accuracy and simplicity, being both lightweight from a computational point of view and easy to use from a user experience perspective. Most notably, it was not the actual distance measurements that mattered, but the trends, i.e., the derivatives, of the contiguous samples that were collected over time once the threshold had been reached.

2.1. System Requirements

Tables 1 and 2 enlist the functional and non-functional requirements, respectively. The functional requirements were classified into user requirements (U), system requirements (S) and error recovery requirements (E). Since the system was divided into two components that interact with each other, i.e., the wearable device and the graphical display, we classified the non-functional requirements into general requirements (G), wearable control unit requirements (W) and user interface requirements (UI).

Table 1. The functional requirements.

	Functional Requirements	Type
FR1	The user must be able to start and stop the system from a user-friendly GUI.	U
FR2	The user must be able to view the data through a user-friendly GUI with a "real time" option.	U
FR3	The user must be able to delete the recorded data.	U
FR4	The user must be able to modify the system settings (e.g., server IP address, the minimum identification height of jumps, the vertical dimension of the graph, etc.).	U
FR5	The system must identify and display when a jump is performed.	S
FR6	The system must provide the user with the number of jumps identified since the beginning of the evaluation.	S
FR7	The system must provide the user with the maximum height reached in the jumps that were performed.	S
FR8	The system must provide correct data, even in the case of several athletes training on the same rink.	S
FR9	The system must not allow the user to modify the data obtained from the sensors.	S
FR10	The system must provide a minimum sample rate of eight times per second when reading data from the sensors.	S
FR11	In case of connection failure with the server IP address entered by the user, the system must continue to execute GET requests.	E

Table 2. The non-functional requirements.

	Non-functional Requirements	Type
NFR1	The athlete must be able to wear the device easily.	G
NFR2	The device must not hinder the athlete's movements or affect them in any way (i.e., weight and position).	G
NFR3	The system must be usable in regular inline figure skating rinks (minimum 40 $\times$ 20 m).	G
NFR4	The control unit must run the Raspbian Lite or compatible operating systems.	W
NFR5	The control unit must communicate via I2C protocol with the sensors and obtain the necessary data with a frequency of less than or equal to three values per second.	W
NFR6	The control unit must grant permission to execute a script and terminate the process corresponding to the user interface.	W
NFR7	The control unit must grant permission to remove a specific file.	W
NFR8	The display of the graphical user interface must be independent of the type of computer used to run it and it must be responsive.	UI
NFR9	The user interface must communicate with the control unit through http requests.	UI
NFR10	The user interface must be run from any device capable of interpreting the PHP language.	UI
NFR11	The user interface must be viewable through a web browser.	UI
NFR12	The user interface must have permission to read and write files locally.	UI
NFR13	The user interface data update frequency must be greater than or equal to three times per second.	UI

2.2. Hardware Components

The system comprised a wearable device that consisted of hardware components that respond to the following critical aspects. The size and weight of the wearable device must not affect athletic movement and must be comfortable for extended training sessions. The accuracy of sensors determines the reliability of the product as it impacts the quality of jump detection, which is the core feature of the system. Finally, the cost is the final important aspect as it affects the possible of uptake the system.

The hardware components that are depicted on the right-hand side of Figure 1 involved the use of the sensors and a single-board computer to perform readings, explain those readings and share them on a local network. The whole system was composed of:

- Raspberry Pi Zero W (https://www.raspberrypi.org/products/raspberry-pi-zero-w/ (accessed on 19 December 2021));
- TCA9548A multiplexer (https://www.ti.com/product/TCA9548A (accessed on 19 December 2021));
- Laser sensor STMicroelectronics VL54L0X (https://www.st.com/en/imaging-and-photonics-solutions/vl53l0x.html (accessed on 19 December 2021));
- MPU-6050 (https://www.radiolocman.com/datasheet/pdf.html?di=102365 (accessed on 19 December 2021));
- UPS-Lite v.1.2 (https://hackaday.io/project/173847-ups-lite (accessed on 19 December 2021)).

The Raspberry Pi Zero W was the single-board computer that we chose because it allows for communication with the sensors through the I2C protocol. The TCA9548A multiplexer was needed for addressing the sensors when retrieving data from them. The VL54L0X laser sensor was chosen for its small size and its accuracy in detecting distances, as well as for its eye safety as a Class-1 laser device that is compliant with the latest IEC 60825-1:2014 standard. The MPU-6050 was the gyroscope–accelerometer sensor that we used to filter the collected data. All sensors were connected to the multiplexer by means of cables and the whole system was powered by a polymer lithium battery through the UPS-Lite v.1.2 board.

2.3. Hardware Architecture and Sensor Placement

Ice Fly skates from EDEA srl (Crocetta del Montello (TV) Italy, https://ice.edeaskates.com/en/ice-skates/ice-fly/ (accessed on 19 December 2021)), a Snow White plate from Double L's (http://www.inlinefigure.com/order_Snow_White.htm (accessed on 19 December 2021)), Zenith wheels (72 mm/78 A) from Roll Line T. M. Technology (Lancenigo (TV), Italy), D-stroyer Outdoor wheels (72 mm/81 A) from Kryptonics Wheels (El Segundo, CA, USA) and a Super professional ambra toe stop by Roll Line T. M. Technology (https://inlineartistic.roll-line.it/it/product/super-professional-ambra/ (accessed on 19 December 2021)) were used for the development of the project and for the tests.

The best position for the infrared sensors was between the second and third wheel, where the ground would be directly visible (see Figure 2). The minimum distance from the ground varied from 5 cm to 10 cm, depending on the type of skate installed under the boot, with its precise value being easily configurable through our software settings.

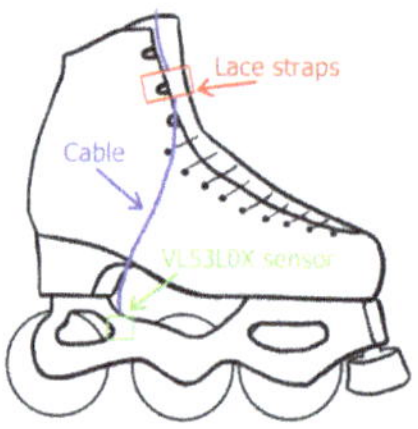

Figure 2. The positioning of the VL53L0X sensor on the boot.

The RPiZW board, multiplexer, gyroscope–accelerometer and battery (control unit) were positioned on the mid-back of the athlete in a central position that would not be significantly affected by falls and would not unbalance the athlete (see Figure 3). In both ice skating and roller/inline skating, the majority of falls are in a forward direction. Skaters attempt to break their falls with their arms or hands in over 90% of falls in both ice skating (93.1%) and roller/inline skating (94.5%) [4]. The *control unit* was worn together with a simple harness made of a stretchy fabric, similar to those used for wearable action cameras.

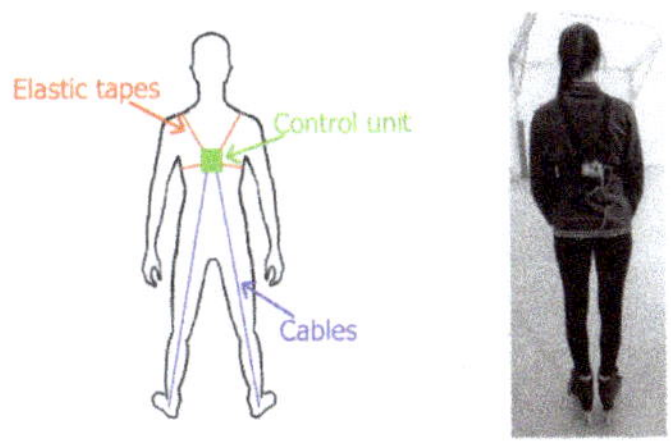

Figure 3. The positioning of the control unit on the athlete.

The resulting system is depicted in Figure 1. The control unit consisted of the RPiZW board, powered by a battery. It communicated with the multiplexer via the I2C protocol through a cable connection, which managed the correct request addressing with the sensors. The I2C protocol provided easy communication without any data loss. It also provided an excellent speed compared to other protocols; I2C uses only two wires for communication so it is lightweight, economical and omnipresent. It also increased the data transfer rate [5]. Inside the control unit, there was also a MPU-6050 sensor that was connected to the multiplexer. The two VL53L0X sensors were connected to the multiplexer outside of the control unit.

2.4. Software Components

The data readings were managed entirely by the RPiZW board described in Section 2.2 (see Figure 1). The RPiZW represented the server of the I2C connection, with the sensors functioning as the client devices. It was, therefore, the RPiZW board that interacted with the sensors and stored the retrieved data in text files after processing them. The RPiZW board also hosted a PHP page: by accessing this page from a device connected to the same network, the user can start or stop the script that manages the communication between the RPiZW board and the sensors and saves results in TXT files.

The implementation was written in Python and was based on the *Adafruit* libraries that were released on GitHub under the name of *circuitpython* (https://github.com/adafruit/circuitpython (accessed on 19 December 2021)), which was a fork of the *micropython* project (https://github.com/micropython/micropython (accessed on 19 December 2021)). The Adafruit project allowed for communicate with sensors over the I2C bus. The Raspberry GPIO libraries (https://pypi.org/project/RPi.GPIO/ (accessed on 19 December 2021)) and Adafruit Blinka (https://pypi.org/project/adafruit-blinka/ (accessed on 19 December 2021)) were also used to communicate with the various pins of the RPiZW board.

Retrieving and storing data supplied by sensors was managed by a Python script executed by the RPiZW board. Its basic idea was similar to that proposed by [6]. The Python script read the values coming from the 3-axis accelerometer and 3-axis gyroscope within the motion sensor and stored them in CSV format. Algorithm 1 provided the implementation of this script in pseudocode. It established an I2C connection with the sensors, waited for their response and saved it in the */var/www/html* path. These files, therefore, contained the measurements for each sensor.

The data management was handled by an external device that was capable of executing PHP code. The task of this device was to execute a PHP page calling a JavaScript script every 3 s. The script executed a GET request through XMLHttpRequest() to retrieve the data from the created file (see Section 2.4) and used it to generate an evaluation chart. This PHP page used HTML code to set the content markup by using Bootstrap 4, JQuery (version: 3.4.1) and Plotly. Settings on the JSON file could be modified by the PHP page executing a POST. Notably, when the settings were changed, the UI triggered a repaint event and the charts were plotted from scratch.

```
Algorithm 1: Sensor readings and sample storage.
import libraries;
i2c bus initialization;
multiplexer initialization;
sensors initialization;
while the process is alive do
    open the results file;
    get sensors data;
    write data in results file;
    close file;
end
```

Another important component was the code for generating the graphics of the page and the request for the script in JavaScript language. Algorithm 2 shows a pseudocode implementation of this. This script was responsible for executing the GET request to retrieve the TXT file generated by the Python script, which was located on the RPiZW board. It then analysed the data and displayed it on a chart using the **Plotly** library. The data analysis was performed considering the height of both feet from the ground. Once the limit threshold was exceeded, if the height of both feet remained above the threshold for at least three consecutive readings, then the series was highlighted by a red point on the graph and was considered a candidate for one single jump. The maximum height was always computed from the lower foot of the two when a candidate jump was detected for each foot.

```
Algorithm 2: Data processing and jump detection.
variables initialization;
data = GET request to the RPiZW ip;
list_of_triples = data splitted;
foreach triple in list_of_triples do
    if both feet are over the ground limit then
        buffer_list.push(triple);
    else if buffer_list.length >= 3 then
        filter false positives with gyroscope data;
        print jump markers;
        print gyroscope data;
        if max_height < jump_height(buffer_list) then
            max_height = jump_height(buffer_list)
        end
        buffer_list.empty()
end
plotChart(list_of_triples)
```

The execution speed of the jump detection script, implementing Algorithm 2, varied according to the number of sensors that were queried. By modifying the script, removing the interrogation part of the MPU6050 sensor and, thus, receiving only the values of the VL53l0X sensors, it was possible to estimate the difference in time taken to complete an overall cycle of data request, data acquisition and storage in a TXT file. The time estimation was performed by printing the timestamp of each iteration. On average, each cycle took about 93.63 milliseconds.

It is reasonable to claim that the system could provide at least 10 measurements per second. By similarly calculating the average time needed to complete a cycle to obtain the values of all three connected sensors, the elapsed time was just above $1/10$ of a second with a difference of 8.13 milliseconds from the previous average. In quadruple jumps, the minimum flying time is 0.64 s (Salchow) and the longest flying time measured is 0.78

s (toe loop); therefore, skaters need more jumping power, especially to perform a quick take-off [7].

The average number of cycles per second of our system was, therefore, satisfactory since it allowed for the reading of at least three measurements during the flying phase, effectively satisfying functional requirement FR10 in Table 1.

2.5. Data Visualisation

Figure 4 shows the web-based graphical user interface that was available to the user. A user could either be a computer operator aiding the athlete during training or the athlete themselves examining the data collected by the system at the end of the training session.

The screenshot shows the application when paused, where the following UI elements appear from top to bottom:

- a [real-time graph] shows the jump heights, sampled by the two laser sensors;
- the ["Number of Jumps"] displays the total number of jumps detected;
- the ["Max Height"] displays the highest out of all detected jumps, with the lowest of the two feet always being reported;
- the ["Gyroscope Samples"] displays the gyroscope values corresponding to the detected jumps.

When the application was in "play" mode, the user could monitor the data being collected in real time and see the graph being plotted with the jump heights sampled by the sensors. When in "pause" mode, the data retrieval stops temporarily, allowing the user to examine the data that was collected up to that moment, magnify the graph and inspect the values more carefully.

Figure 4. The web-based GUI. The graph shows jump heights sampled in real time by the sensors.

2.6. Experimentation on Human Subjects

We conducted multiple series of tests involving four humans subjects, two male and two female, who provided and signed an informed consent document. Two of them are high profile athletes in the discipline of inline figure skating, boasting a world-class career; the other two are athletes of regional and national level.

Each subject involved in the experiments was informed about the nature of the research and explicitly agreed for the research team to collect data while they performed jumps, technical elements or other athletic exercises.

All experiments were conducted in a gym under the supervision of a senior member of the research team and a senior athlete in the discipline of inline figure skating. The research presented in this paper and the experimentation involving human subjects were approved by the Ethical Committee of the Ca' Foscari University of Venice.

2.6.1. Experiment #1: Squat Jumps

For the first experiment, our subjects performed a series of 10 squat jumps each, with the purpose of evaluating the accuracy of the two laser sensors located under each foot. The squat jumps were executed wearing skates and from a standing position, using both feet to jump as high as possible.

As a reference for our measurements, we adopted an Abalakov belt [8,9]: a standard practice in sports involving jumps, which we discuss in detail in Section 5.3. It essentially consists of a rope bound to the athlete's belt that is free to pull out during the vertical motion performed by the athlete. Once the athlete lands, the rope does not retract; hence, the amount of rope that pulled out equals the vertical distance covered during the jump.

2.6.2. Experiment #2: Technical Element Detection

A second series of tests were conducted to evaluate the effectiveness of the technical element detection algorithm, which is the core feature of the proposed system. Our subjects performed a series of different jumps and technical elements on a regular inline figure skating rink (20×40 m^2) with a quartz concrete surface.

As detailed in the Section 2.3, the materials used in our tests were:

Skating boots: Ice Fly (athletes 1, 3 and 4) and Chorus (athlete 2) from EDEA srl;

Frames: Snow White from Double L's (all athletes);

Wheels: Speedmax from Double L's (athletes 1, 2, 3); Zenith (72 mm/78 A) from Roll Line T. M. Technology (second and third wheels) and D-stroyer Outdoor (72 mm/81 A) from Kryptonics Wheels (first wheel) (athlete 4);

Toe stops: Ambra from Roll Line T. M. Technology (all athletes).

This equipment was provided by the research team in order to make tests as much consistent as possible.

3. Results

In this section, we report the results of our experiments and show the collected data and statistics.

3.1. Results of Experiment #1: Squat Jumps

Table 3 reports the following statistics of the squat jumps, conducted in a series of 10, for each athlete: the minimum, maximum and average heights $H_{min/max/\mu}$, read by the laser sensors; the average value H_μ^{ref} of heights, measured using the Abalakov belt; the minimum, maximum and average absolute deviation $D_{min/max/\mu}$ between the values read by the sensors and those measured by the belt; the variance σ_D^2 of the deviations, with respect to the average deviation D_μ.

In this experiment, accuracy was defined as the ratio between the average absolute deviation D_μ and the average reference height H_μ^{ref}:

$$Accuracy = 1 - \frac{D_{\mu}}{H_{\mu}^{ref}}$$

In Figure 5, we provide a box plot of the deviations between the values read by the laser sensors and those measured by the Abalakov belt, used as a reference, for each athlete.

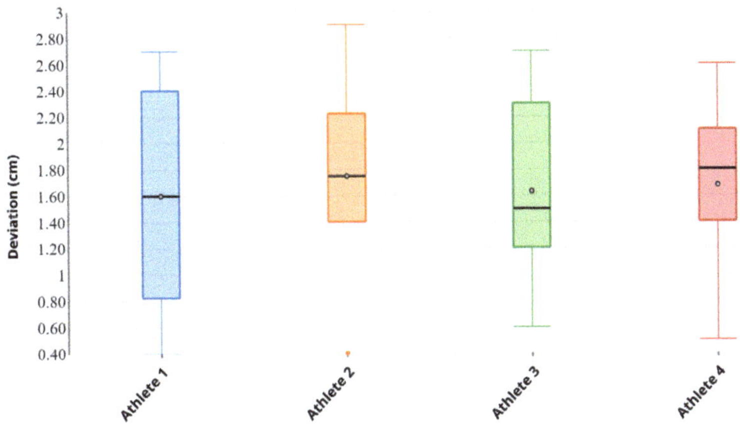

Figure 5. A box plot of the deviations between the sensor data and the belt reference measurement for each athlete. Empty dots represent the mean and black segments are the median. Despite the different physical characteristics of the athletes, the deviations were quite stable, with an overall mean of around 1.61 cm and a relatively low variance.

Table 3. Squat jump measurements: the min/max/avg heights read from the sensors; the average reference height read from the Abalakov belt; the min/max/avg deviation between the data from the sensors and the belt; the variance of the deviations; accuracy.

Athlete #	H_{min} (cm)	H_{max} (cm)	H_{μ} (cm)	H_{μ}^{ref} (cm)	D_{min} (cm)	D_{max} (cm)	D_{μ} (cm)	σ_D^2	Accuracy
1	20.3	23.6	22.11	23.71	0.4	2.7	1.60	0.73	93.25%
2	28.6	32.4	30.39	32.14	0.1	2.9	1.75	0.65	94.56%
3	25.3	29.6	27.70	26.07	0.6	2.7	1.63	0.51	93.74%
4	38.4	42.1	40.43	38.75	0.5	2.6	1.68	0.46	95.66%

3.2. *Results of Experiment #2: Technical Element Detection*

Algorithm 2 in Section 2.4 describes the technical element detection system as a three-stage pipeline implemented through a series of nested conditional computations. Table 4 shows the intermediate results for each of the three stages, testing the internals of Algorithm 2 on a series of different technical elements performed by the athlete subjects. Each column of the table represents a stage as a Boolean value.

Has3Triggers: This indicates the first stage of detection; when the VL53L0X laser sensor detected three consecutive peaks of height above the threshold, a candidate jump was considered;

IsCapped: When the Has3Triggers stage succeeded, up to five additional samples were taken into consideration for detecting the landing point of the candidate jump. Then, the values read by the MPU-6050 gyroscope were processed. When the values equalled the hard limit value 250 (refer to the appendix for the details of this hard limit), it meant the athlete was spinning too quickly and hence, that it was not a jump;

IsJump: The rightmost column presents the ultimate result of a Boolean value indicating whether the given element was detected as a valid jump or not.

Table 5 reports the accuracy of the technical element detection at different thresholds for each athlete. These data involve non-vertical jumps while the athlete moved and performed complex spins and figures. The thresholds represent the minimum height at which our system considered a jump as a candidate technical element to be processed by the algorithm. The different physical characteristics and jumping capabilities exhibited by the athletes had an impact on the choice of threshold.

Table 4. Intermediate Boolean states of the three-stage pipeline for the detection of technical elements implemented by Algorithm 2.

Element	Has3Triggers	IsCapped	IsJump
Push	No	No	No
Upright Spin	No	No	No
Sit Spin	No	No	No
Lunges Forward	No	No	No
Waltz Jump	Yes	No	Yes
Single Loop	Yes	No	Yes
Single Axel	Yes	No	Yes
Double Salchow	Yes	No	Yes
Double Toe Loop	Yes	No	Yes
Double Loop	Yes	No	Yes
Double Flip	Yes	No	Yes
Double Lutz	Yes	No	Yes
Flying Upright Spin	Yes	Yes	No
Flying Sit Spin	Yes	Yes	No
Flying Camel Spin	Yes	Yes	No
Flying Change Sit Spin	Yes	Yes	No

Table 5. The accuracy of the detection of different technical elements for multiple thresholds and for each athlete. The thresholds played a significant role in delivering optimal accuracy and were essentially dependent on the characteristics of each athlete, such as height, weight and strength.

Technical Element	Athlete #	Treshold						
		10 cm	**15 cm**	**18 cm**	**20 cm**	**25 cm**	**30 cm**	**35 cm**
Double Salchow		100%	100%	100%	100%	85%	50%	30%
	1	100%	100%	100%	100%	40%	0%	0%
	2	100%	100%	100%	100%	100%	60%	20%
	3	100%	100%	100%	100%	100%	40%	0%
	4	100%	100%	100%	100%	100%	100%	100%
Double Toe Loop		100%	100%	100%	100%	85%	50%	30%
	1	100%	100%	100%	100%	40%	0%	0%
	2	100%	100%	100%	100%	100%	60%	20%
	3	100%	100%	100%	100%	100%	40%	0%
	4	100%	100%	100%	100%	100%	100%	100%
Flying Camel Spin		0%	0%	12.5%	55%	100%	100%	100%
	1	0%	0%	40%	100%	100%	100%	100%
	2	0%	0%	0%	20%	100%	100%	100%
	3	0%	0%	10%	80%	100%	100%	100%
	4	0%	0%	0%	20%	100%	100%	100%
Combination Spin		100%	100%	100%	100%	100%	100%	100%
	1	100%	100%	100%	100%	100%	100%	100%
	2	100%	100%	100%	100%	100%	100%	100%
	3	100%	100%	100%	100%	100%	100%	100%
	4	100%	100%	100%	100%	100%	100%	100%
Step Sequence		0%	30%	95%	100%	100%	100%	100%
	1	0%	100%	100%	100%	100%	100%	100%
	2	0%	0%	100%	100%	100%	100%	100%
	3	0%	20%	100%	100%	100%	100%	100%
	4	0%	0%	80%	100%	100%	100%	100%

4. Discussion

In this section, we discuss our experiments and comment on the data collected therein and evaluate the effectiveness of the proposed system.

4.1. Discussion of Experiment #1: Squat Jumps

From data shown in Table 3, one relevant piece of information emerges: subjects of different gender, professional level, athletic performance and physical characteristics produced the same results in terms of accuracy, thus accuracy was not related to jump height. Secondly, Figure 5 highlights the overall low variance across the subjects, implying that the laser sensors were reliable in most scenarios.

4.2. Discussion of Experiment #2: Technical Element Detection

The data collected during the second experiment suggest that using $N = 3$ as the number of contiguous samples above the threshold needed to trigger a valid jump (as in Definition 2) was a good trade-off. As a matter of fact, sampling heights 10 times per second allowed for three contiguous samples to cover a time span of 0.78 s, which is the average duration of a typical jump according to [7].

The first two technical elements, a double Salchow (see Figure 6) and a double toe loop, are essentially simple jumps and in both cases, our system always detected them with a threshold empirically calibrated to 20 cm. As the threshold increased, the jumps performed by athletes featuring the lowest average height were filtered out, i.e., not detected as valid elements. It is important to remark that three consecutive samples beyond the threshold had to be read by the software in order to consider it as a valid element. In some cases, the system did not detect a jump where the athlete surpassed the threshold for just 2 samples as a valid element.

Figure 6. Double Salchow. Max height detected: 28.8 cm.

The flying camel spin (see Figure 7) is a spin with flying entrance, thus a jump takes place before starting the rotation on the floor. The system detected the jump entrance as an element, which was not supposed to happen and represented an error. Considering that the goal of a spin is not to achieve maximum height, the entrance jump reached a maximum height that was lower than that of an element based on a plain jump; hence, the number of errors decreased as the threshold increased. In order to avoid errors at lower threshold values, we introduced an additional filtering condition to the detection algorithm (see Algorithm 2 in Section 2.4).

Figure 7. Flying camel spin. Max height detected: 23.6 cm. The element was discarded by the algorithm as it detected a fast spin after landing.

Combination spins require one foot to remain on the floor. The obvious consequence of this was no incorrect detections arose with any threshold value because this element technically does not constitute a jump for our system, since both feet must pass the threshold for three samples for this to occur.

As far as a step sequence is concerned, the athlete has to perform difficult turns and steps and even choreographic elements, such as little hops or jumps. Only athlete #1 did not execute any jumps, even small ones, within the step sequence; hence, with thresholds higher than 15 cm, our system did not detect any jumps during the recording.

In all cases where the threshold was set to 10 cm, at least one element based on a jump was always incorrectly detected during the recording of the step sequence. The reason behind this was that the athletes might perform special movements where their feet slant with respect to the floor, causing the laser sensors to measure a further distance than the threshold due to the angle, especially when the threshold was set very low.

When setting the threshold at 20 cm, we achieved the best probability of success, which was increased even more by the algorithm filtering out false positives, such as spin jumps, through the control of the gyroscope values.

4.3. *Accuracy of Jump Height Measurements*

A mistake that an athlete can make when performing a technical element is to bend the legs or ankle during the jump; this caused the laser sensor installed under each skate to read a distance from the ground that was greater than the actual vertical height. Figure 8 shows an athlete bending legs in such a way that significantly impacted the measurements of vertical height due to tilting.

We analysed 40 videos (10 for each of the four subjects) that were recorded during our test sessions and, for each jump occurring in the performance, we extracted the angle between the foot and the ground by measuring it on the specific frame in which the highest tilt value appears, i.e., we picked the worst cases. In order to obtain an acceptable result, we only took into account frames where the athlete's orientation was favourable for measuring such an angle, meaning that the third-dimension component was negligible.

With reference to Figure 8, we let $\overline{OA}$ be the distance from the ground as read by the sensor, $\overline{OB}$ be the vertical height and $\widehat{AOB}$ be the tilt angle; then the equation $\overline{OB} = \overline{OA}\cos(\widehat{AOB})$ held.

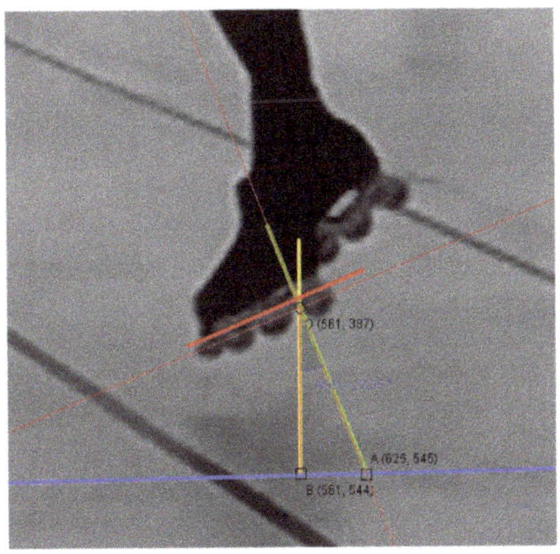

Figure 8. A sample calculation of vertical height $\overline{OB}$ given the tilt angle $\widehat{AOB}$ and the distance from the ground $\overline{OA}$.

Table 6 shows the vertical heights $\overline{OB}$ calculated for each combination of the minimum, maximum and average angles $\widehat{AOB}$ and the minimum, maximum and average distances from the ground $\overline{OA}$, yielding nine results in total. The last column shows the accuracy, here defined as the ratio between the value $\overline{OA}$ measured by the laser sensor (the green hypotenuse of the triangle in Figure 6) and the actual vertical distance from the ground $\overline{OB}$ (the yellow vertical cathetus of the triangle in Figure 6). In other words, the accuracy was equal to the cosine $cos(\widehat{AOB})$ of the tilt angle.

Figure 9 depicts the box plot of the tilt angle data extracted from the video recordings, displaying the relatively low variance and a mean of around 8°. As a general note, the system delivered a high accuracy overall: the mean angle (8°) provided 99% accuracy; the upper bound of the third quartile (10°) provided 98%; and the minimum angle (2°) had an accuracy of 99%. Even the outliers that do not belong in the box plot in Figure 9 provided a good accuracy, such as the worst case of $\widehat{AOB}_{out} = 30°$.

Table 6. The vertical heights $\overline{OB}$ calculated for each combination of the min/max/avg distances $\overline{OA}$ and angles $\widehat{AOB}$.

	$\overline{OA}_{min}$ = 20.3 cm	$\overline{OA}_{max}$ = 42.1 cm	$\overline{OA}_{\mu}$ = 30.2 cm	Accuracy
$\widehat{AOB}_{min} = 2°$	$\overline{OB}$ = 20.3 cm	$\overline{OB}$ = 42.1 cm	$\overline{OB}$ = 30.2 cm	99%
$\widehat{AOB}_{\mu} = 8°$	$\overline{OB}$ = 20.1 cm	$\overline{OB}$ = 41.7 cm	$\overline{OB}$ = 29.9 cm	99%
$\widehat{AOB}_{max} = 15°$	$\overline{OB}$ = 19.7 cm	$\overline{OB}$ = 40.8 cm	$\overline{OB}$ = 29.3 cm	97%
$\widehat{AOB}_{out} = 30°$	$\overline{OB}$ = 17.6 cm	$\overline{OB}$ = 36.5 cm	$\overline{OB}$ = 26.2 cm	87%

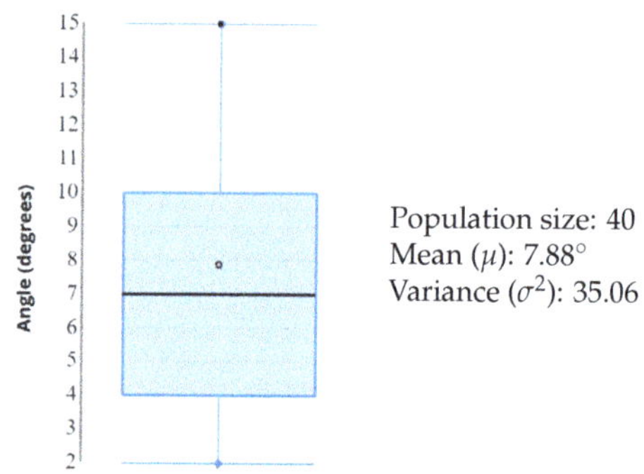

Figure 9. The statistics box plot of the 40 tilt angles measured from the video frames. The empty dot represents the mean and the black segment represents the median.

In order to analyse the variability of the angles in our population, we modelled a beta distribution using the maximum likelihood method [10] for determining the α and β parameters. The domain of the beta distribution was rescaled into the $(0°, 90°)$ interval to match the whole possible range of the angular data, where 0° represented the perfectly vertical orientation of the laser sensor with respect to the ground and 90° represented the maximum absolute tilt. Figure 10 depicts the estimated density function and the cumulative distribution function (CDF) of this inferential model, showing, for instance, that an athlete's foot exhibited a tilt angle lower than 20° with roughly a 90% probability, hence an accuracy lower than $cos(20°) = 0.93962 \approx 94\%$ with a 10% probability.

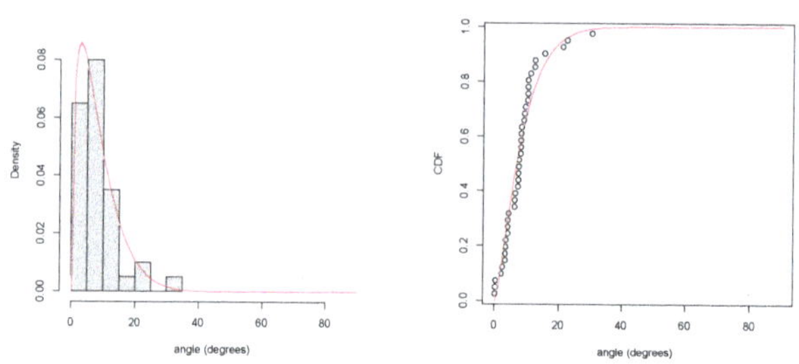

Figure 10. Inferential statistics on the population of tilt angles: the estimated density function in the range 0–90 degrees (**left**) and the cumulative distribution function (**right**).

4.4. In-Depth Analysis of a Jump

We now delve into a detailed example of technical element detection consisting of a non-trivial jump. Figure 11 shows the data from the two laser sensors (blue and red) and the three axes of the gyroscope (green, purple, azure) during the execution of a double Lutz (upper chart) and a flying sit spin with foot change (lower chart). The hard limit value (250) reached by the various axes of the gyroscope (B) occurred when the athlete was in the air (A) during the rotation phase of the jump. Once landed, the gyroscope values dropped considerably. In the case of spins with a flying entrance, the data instead showed the persistent reaching of the hard limit of the gyroscope in each axis after the landing phase (D) and not just in correspondence with the flight phase (C).

It was, therefore, possible to reject false positives (flying spins) by performing an evaluation of the gyroscope data after the landing. The same differences were also observed in comparisons between the jumps and spins performed in the tests. Each test was performed three times, and all jumps were accurately detected with a properly tuned threshold value (20 cm). The harness did not disconnect during the tests and the device did not suffer any damage and remained bound to the harness without creating any difficulties in the execution of the elements.

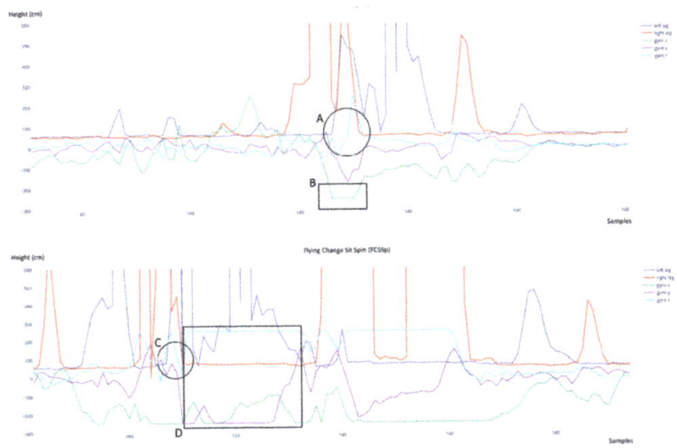

Figure 11. A comparison between a double Lutz (**upper chart**) and a flying sit spin with foot change (**lower chart**). The gyroscope hard limit was reached while jumping (the cyan line in the B area above) rather than after landing (the cyan line in the D area below).

Figure 12 depicts the chart related to three double Lutz jumps that were performed during the tests. The values read by the laser sensors mounted under the left and right skate are highlighted in green and blue, respectively. The red dots indicate that both feet exceeded the threshold: after three consecutive occurrences, the system identified the sequence as a valid jump. The graph in Figure 12 shows three groups of three red dots, meaning that all three double Lutz jumps were properly detected.

Figure 13 zooms in on the rightmost part of the graph in Figure 12, showing the last double Lutz of the series of three in detail. The first peak below 200 mm (both in blue and green) represents the starting phase of the execution of the technical element. The following blue peak over 200 mm represents the take-off phase, the in air phase and the landing phase of the jump. At the same time, the green peak represents the left leg rising slightly more than the right as it helped the body in balancing its weight.

Finally, the landing phase was performed on the right foot only, since the blue line flattens much earlier compared to the green line. The green line then reached the hard limit of the sensor at 8190 mm, meaning that the left foot was so tilted that the sensor did not identify the floor: this represented the exit phase, where the left foot was indeed positioned parallel to the ground. However, since the right foot was on the ground, the system did not evaluate the exit phase as a jump, which was the correct behaviour.

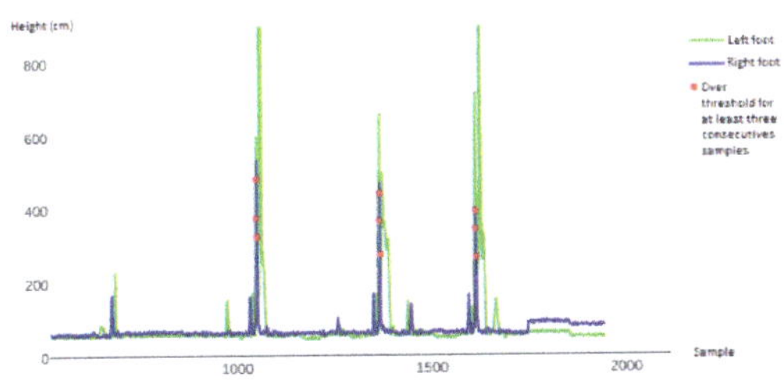

Figure 12. The green and blue lines relate to the left and right foot, respectively. The red dots highlight when both feet surpassed the threshold, with three red dots indicating that a valid jump was detected. This chart shows three successful detections of three technical elements (double Lutz).

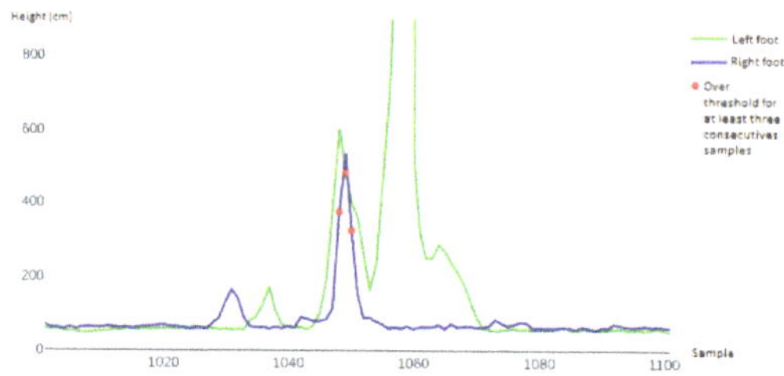

Figure 13. A zoom-in of the third (rightmost) double Lutz of the series of three in Figure 12. After the flying phase, marked by the three red dots and regularly detected as a valid jump, the left foot went parallel to the ground, hence the high green peak reaching the hard limit of the laser sensor, whereas the right foot (blue line) remained on the ground. This scenario corresponded to the exit phase of the element and was not detected as a jump.

5. Related Work

In this section, we discuss the most relevant papers in the literature that are related to our contribution. Distinguishing between cloud-based proposals and measurement systems, we compared our solution to a series of related tools that were, for the most part, not designed for inline skating. Almost no literature seems to exist on the discipline that this paper targeted, i.e., inline figure skating, although proposals worth a comparison do exist for similar sports, such as skating on ice, or for technologies based on similar principles.

5.1. Cloud Systems and BLE

Cloud systems, such as that proposed in [11], make data accessible everywhere, but also require a Wi-Fi connection, to enable the RPiZW board to access the Internet, or a GSM (Global System for Mobile Communications) module with a SIM (Subscriber Identity Module) to be added to the device. That is arguably a limitation in some scenarios. Our system does not depend on an Internet connection and can, therefore, be used anywhere.

A crucial aspect is the cost of time for uploading data from the wearable device to the processing device (a computer or a server). It requires processing time on the RPiWZ side and can, therefore, increase the time elapsed between contiguous sensor reads. A solution that does not need an Internet connection and uses BLE to transmit data from the sensors to a device, such as a smartphone, is presented in [12], where a mobile device manages the whole cloud upload process autonomously. Their system does not need a RPiZW board and resolves the constraints of a wired system, although it requires a BLE card, which increases the size of the two sensors (one for each foot). Our laser sensor is small enough to be placed in between the skate wheels, though adding a Bluetooth transmitter and a battery would make it impossible to fit.

5.2. Number of IMU Sensors

Several proposals in the literature use just one IMU sensor for each foot, for example in [13], whereas others take advantage of multiple IMU sensors.

In [14], up to 13 IMUs are adopted and the system is able to capture 6-degrees-of-freedom full-body motion in real time as long as there is at least one motionless foot at any given time. Generally, using multiple sensors placed in different key positions of the body allows for more precise measurements and more advanced results than simply detecting jumps and estimating their height. The ability to reproduce body movements in a 3D model would be more informational to the end user compared to our UI, but it would arguably rely on too many sensors, each powered by a battery and connected to a microcontroller that would use a BLE card to transmit the data to the server. Such a setup would be necessarily less lightweight and less comfortable than a system with fewer sensors, such as ours. Moreover, the IMU sensor that our system adopts is placed on the athlete's back, both for safety reasons and because that placement is best for detecting body rotation [4].

The system proposed in [15], despite addressing ice skating rather than inline figure skating, is worth comparison due to method in which the measurements are carried out. Jump detection is based on waist-mounted IMUs and measures both the height and the rotation angle rather than only the height, as in our system. The height measurements take place at the peak of a jump, as with our algorithm, although the sampled values are subject to an error ranging between 3.3 cm and 7.81 cm, which is higher, on average, than ours. That paper also compares different algorithms for detecting valid vs. invalid jumps, whereas our system is based on one single algorithm tailored for yielding optimal results. Table 5 shows that the hit rate of our jump detection algorithm seems to be generally higher, assuming that a fine-tuned threshold is configured. The algorithms proposed in [15] depend on a few parameters that need to be manually configured; therefore, the paper does not provide a generic solution that adapts to all scenarios, whereas our system relies on one single parameter—the threshold.

5.3. Comparison to Other Systems

Many systems exist in the literature for measuring the different characteristics of a jump that are not related to skating as a sport. Most of these systems focus on providing the maximum height of the jump and collecting data for other purposes, although some provide a good amount of detail, such as the number of jumps performed, the intensity of the jumps and statistics related to a medical evaluation of athletic performance that is aimed at avoiding injuries.

- Sargent jump test [16,17];
- Abalakov jump test [8,9];
- Conductance platform [18,19];
- Insoles for pressure measurement: Pedar (https://www.novel.de/products/pedar/ (accessed on 19 December 2021));
- Optical detection system: Optojump (http://www.optojump.com/Cos-e-Optojump.aspx (accessed on 19 December 2021)) [18,19];
- Video triangulation system: TRACAB Optical (https://tracab.com/products/tracab-technologies/tracab-optical (accessed on 19 December 2021));
- Inertial measurement unit: XSENS (https://www.xsens.com/products/mvn-analyze (accessed on 19 December 2021)), VERT (https://www.myvert.com/ (accessed on 19 December 2021)) [20].

These systems feature useful aspects but none of them fully coincide with the design goals and requirements of our system, since they were not designed for the specific purpose of detecting jumps in figure skating. The Sargent jump test, Abalakov jump test, conductance platform and Optojump all require the jump to be performed within the measurement area, which is a major limitation. Despite Garcia-Lopez et al. [8] and Optojump offering a better overall precision compared to our system by measuring jumps through accurately detecting the flying curve performed by athletes, our system still detects technical elements with a high degree of reliability without imposing the above-mentioned space limitations and with a less expensive technology. On the other hand, the insoles for pressure measurement are not influenced by a measurement area though they present other issues, such as the size and weight of the instrumentation: the data analyser of the Pedar Insoles alone has a size of 15 × 10 × 4 cm and a weight of 400 g.

Analysing data from sensors is also required to determine when a jump, or another element, is occurring. Some products, such as the devices by the company VERT, are specialised in tracking the execution of jumps, while those by XSENS allow the reconstruction of the athlete's movements and then animate a 3D model. Our proposal instead aims at detecting when a jump is performed, excluding any other elements performed by the athlete. Certain movements, such as flexing the knees and quickly extending them back to the previous position, could cause evaluation errors in systems that use IMU technology with few sensors for measurement. On the other hand, placing a lot of sensors in multiple key points of the body is not comfortable for the athlete and the cost of the resulting system would be pretty high, whereas our solution aims to reduce the number of sensors to the minimum whilst maintaining a high accuracy threshold and an affordable cost.

Out of the systems aimed at the competitive field, the best is arguably video triangulation. The major limitations of this technology, however, are the high cost of the equipment and the fact that it is highly conditioned by the position of the cameras, the type of background, the lighting of the environment and obstacles, including any other athletes in the same shot.

6. Conclusions

The system proposed in this paper is a hardware–software solution designed as a training aid for athletes in the discipline of inline figure skating. It is capable of interpreting data from non-invasive wearable sensors and detecting jumps and other elements executed by the athlete, ultimately providing a comprehensive view and analysis of their performance.

The sensors are installed under both skates worn on each foot and on the back of the athlete. The vertical distance measured by those sensors had an overall accuracy of 97% to 99% when the athlete does not bend their feet much while jumping, and even in the worst cases, an accuracy of over 85% was evaluated. The effectiveness of the jump detection system relies on these measurements and was proven to be 95% to almost 100% accurate, once a proper threshold was set. The threshold is a statically configurable value that represents the minimum distance from the floor that the system considers to be a jump and needs to be manually set by the user according to a number of physical parameters of the athlete, such as height, mass and strength. The experiments showed that a threshold of 20 cm yields good results for most athletes, despite their gender, skill and strength, as the results in Table 5 demonstrate. Furthermore, data from the gyroscope sensor is used to filter out false positives in cases such as flying spins, further increasing the overall quality of the detection.

If our results were to be confirmed by further tests conducted with a greater number of athletes and over a wider range of environments, locations, floor materials etc., that would arguably make the proposed system better than most jump detection systems available in the literature, especially those relying on video triangulation or on IMU wearable devices.

Among the major advantages of our system, it is capable of carrying out measurements with no space limitations and without depending on the strength of the Wi-Fi signal, which is usually an acceptable constraint in most cases as a Wi-Fi signal is easily extendable by installing signal repeaters that was readily available on the market. It also operates without a Wi-Fi connection once the acquisition script has been run, leaving data analysis for a later time.

The system has proven to be effective within training contexts where the counting and analysing of jumps and other technical elements is useful for a number of reasons, from improving overall athlete performance to fixing mistakes in execution.

Future Challenges

Among the limitations of this system, its physical size is an open issue; a future upgrade could address this by crafting a custom case or box, whereas the current prototype was built by assembling consumer products that were already available on the market. We believe it is possible to greatly reduce the size of the device by carefully selecting finer (and more expensive) hardware. Removing the wires between the sensors and the RPiZW and adopting a fully wireless communication between the laser sensors and the RPiZW board would also be an important improvement as far as ergonomics are concerned. That would require a battery-powered communication board to be installed on the back of the athlete, rendering further miniaturisation all the more necessary.

Author Contributions: Conceptualization, A.C. and A.P.; methodology, A.C. and A.S.; software, A.P. and A.S.; validation, A.S.; data curation, A.P. and A.S.; writing—original draft, A.S. and A.P.; writing—revisions and editing, A.S. and A.C. All authors have read and agreed to the published version of the manuscript.

Funding: This research received no external funding.

Institutional Review Board Statement: The study was approved by the Ethics Committee of Ca' Foscari University (01/13/2022).

Informed Consent Statement: Informed consent was obtained from all subjects involved in the study.

Data Availability Statement: Data available in a publicly accessible repository that does not issue DOIs. Publicly available datasets were analyzed in this study. This data can be found here: https://github.com/alvisespano/sensors_2022_skate.

Acknowledgments: We would like to thank the anonymous reviewers for their suggestions to improve the quality of this paper, and also Carlo Gaetan for supporting us in the statistical processing of the collected data.

Conflicts of Interest: The authors declare no conflict of interest.

Abbreviations

The following abbreviations are used in this manuscript:

AWG	American Wire Gauge
BLE	Bluetooth Low Energy
FDM	Fused Deposition Modelling
FIRS	Fédération International de Roller Sports
GND	Ground, Common Reference Potential of Value 0 V
GPIO	General Purpose Input/Output
I2C	Inter Integrated Circuit
IoT	Internet of Things
IIoT	Industrial Internet of Things
IMU	Inertial Measurement Unit
ISU	International Skating Union
JST-XH	Japan Solderless Terminal
PLA	Polylactic Acid
RPiZW	Raspberry Pi Zero W
SCL	Serial Clock
SDA	Serial Data
SOV	Scale of Values
VCC	Positive Power Supply
WS	World Skate

Appendix A. The Hard Limit of Gyroscope Values

Each 16-bit gyroscope measurement has a full scale defined in FS_SEL (Register 27). For each full-scale setting, the gyroscope sensitivity per LSB in GYRO_xOUT is shown in the table below (Source: https://pdf1.alldatasheet.com/datasheet-pdf/view/1132807/TDK/MPU-6050.html (accessed on 19 December 2021)):

Table A1. Gyroscope sensitivity.

FS_SEL	Full Scale Range	LSB Sensitivity
0	±250°/s	131 LSB/°/s
1	±500°/s	65.5 LSB/°/s
2	±1000°/s	32.8 LSB/°/s
3	±2000°/s	16.4 LSB/°/s

References

1. Rinehart, R.E. *Inline Skating in Contemporary Sport: An Examination of Its Growth and Development*; University of Waikato: Hamilton, New Zealand, 2013.
2. International Skating Union. Special Regulations and Technical Rules; International Skating Union Website. 2021. Available online: https://www.isu.org/figure-skating/rules/fsk-regulations-rules/file (accessed on 19 December 2021).
3. International Skating Union. Technical Panel Handbook—Single Skating; International Skating Union Website. 2021. Available online: https://www.isu.org/figure-skating/rules/sandp-handbooks-faq (accessed on 19 December 2021).
4. Knox, C.L.; Comstock, R.D. Video analysis of falls experienced by paediatric iceskaters and roller/inline skaters. *Br. J. Sport. Med.* **2006**, *40*, 268–271. [CrossRef] [PubMed]
5. Mankar, J.; Darode, C.; Trivedi, K.; Kanoje, M.; Shahare, P. Review of I2C protocol. *Int. J. Res. Advent Technol.* **2014**, 2, 474-479.
6. Tabbakha, N.E.; Tan, W.H.; Ooi, C.P. Indoor location and motion tracking system for elderly assisted living home. In Proceedings of the 2017 International Conference on Robotics, Automation and Sciences (ICORAS), Melaka, Malaysia, 27–29 November 2017; pp. 1–4.
7. Knoll, K.; Härtel, T. Biomechanical Conditions for Stabilizing Quadruple Figure Skating Jumps as a Process of Optimization. In Proceedings of the ISBS-Conference Proceedings Archive, Beijing, China, 22–27 August 2005.
8. Garcia-Lopez, J.; Peleteiro, J.; Rodgriguez-Marroyo, J.; Morante, J.; Herrero, J.; Villa, J. The validation of a new method that measures contact and flight times during vertical jump. *Int. J. Sport. Med.* **2005**, *26*, 294–302. [CrossRef] [PubMed]
9. Rodríguez-Rosell, D.; Mora-Custodio, R.; Franco-Márquez, F.; Yáñez-García, J.M.; González-Badillo, J.J. Traditional vs. sport-specific vertical jump tests: Reliability, validity, and relationship with the legs strength and sprint performance in adult and teen soccer and basketball players. *J. Strength Cond. Res.* **2017**, *31*, 196–206. [CrossRef] [PubMed]
10. Rossi, R. *Mathematical Statistics: An Introduction to Likelihood Based Inference*; John Wiley & Sons: Hoboken, NJ, USA, 2018; pp. 1–422. [CrossRef]
11. Qureshi, F.; Krishnan, S. Wearable hardware design for the internet of medical things (IoMT). *Sensors* **2018**, *18*, 3812. [CrossRef] [PubMed]
12. Mannini, A.; Genovese, V.; Sabatini, A.M. Online decoding of hidden Markov models for gait event detection using foot-mounted gyroscopes. *IEEE J. Biomed. Health Inform.* **2013**, *18*, 1122–1130. [CrossRef] [PubMed]
13. Niu, X.; Li, Y.; Kuang, J.; Zhang, P. Data fusion of dual foot-mounted IMU for pedestrian navigation. *IEEE Sens. J.* **2019**, *19*, 4577–4584. [CrossRef]
14. Zheng, Y.; Chan, K.C.; Wang, C.C. Pedalvatar: An IMU-based real-time body motion capture system using foot rooted kinematic model. In Proceedings of the 2014 IEEE/RSJ International Conference on Intelligent Robots and Systems, Chicago, IL, USA, 14–18 September 2014; pp. 4130–4135.
15. Bruening, D.A.; Reynolds, R.E.; Adair, C.W.; Zapalo, P.; Ridge, S.T. A sport-specific wearable jump monitor for figure skating. *PLoS ONE* **2018**, *13*, e0206162. [CrossRef] [PubMed]
16. Palaniappan, B.; Pasupatham, V.; Kalirathinam, D. Effect of Static Stretching on Vertical Jump Performance on Apparently Healthy Subjects. *IOSR J. Nurs. Health Sci.* **2013**, *2*, 50–52. [CrossRef]
17. Noreddine, Z.; Djamel, M.; Houcine, B.; Mohammed, S. The Effect of the Contrastive Training Using Weights and Plyometrics on the Development of the Vertical Jump Ability to Improve the Performance of the Smash for Volleyball Players. *Eur. J. Phys. Educ. Sport* **2016**, 24–30. [CrossRef]
18. García-López, J.; Morante, J.C.; Ogueta-Alday, A.; Rodríguez-Marroyo, J.A. The type of mat (contact vs. photocell) affects vertical jump height estimated from flight time. *J. Strength Cond. Res.* **2013**, *27*, 1162–1167. [CrossRef] [PubMed]
19. Santos-Lozano, A.; Gascón, R.; López, I.; Garatachea-Vallejo, N. Comparison of two systems designed to measure vertical jump height. *Rev. Int. Cienc. Deporte* **2014**, *10*, 123–130. [CrossRef]
20. Benbasat, A.Y.; Paradiso, J.A. An inertial measurement framework for gesture recognition and applications. In *International Gesture Workshop*; Springer: Berlin, Germany, 2001; pp. 9–20.

sensors

MDPI

Article

Magnitude and Shape of the Forces Applied on the Foot Rest and Paddle by Elite Kayakers

Pedro Bonito [1], Miguel Sousa [2], Fernando José Ferreira [2], Jorge Fonseca Justo [2] and Beatriz Branquinho Gomes [1,3,*]

1 Faculty of Sport Sciences and Physical Education, University of Coimbra, 3040 Coimbra, Portugal; pedro_a_bonito@hotmail.com
2 Department of Mechanical Engineering, School of Engineering, Polytechnic of Porto, ISEP/IPP, 4200 Porto, Portugal; miguelfbsousa@yahoo.com (M.S.); fjf@isep.ipp.pt (F.J.F.); jfj@isep.ipp.pt (J.F.J.)
3 Research Unit for Sport and Physical Activity—CIDAF (uid/dtp/042143/2020), University of Coimbra, 3040 Coimbra, Portugal
* Correspondence: beatrizgomes@fcdef.uc.pt; Tel.: +351-(239)802770

Abstract: The study aimed to investigate the magnitude and shape of the forces applied on the foot rest, foot strap, and paddle. Thirteen elite male kayakers participated in this study and performed a 2-min test simulating 500 m race pace in a kayak ergometer. Forces applied by the kayakers on the paddle, foot rest, and foot strap were measured with load cells and recorded by an electronic measuring system. The magnitude of the peak forces applied on the foot rest (left: 543.27 ± 85.93; right: 524.39 ± 88.36) approximately doubled the ones applied on the paddle (left: 236.37 ± 19.32; right: 243.92 ± 28.89). The forces on the foot strap were similar in magnitude to the paddle forces (left: 240.09 ± 74.92; right: 231.05 ± 52.01). A positive correlation was found between the peak forces applied on the foot rest and paddle on the same side ($p < 0.001$). When comparing the best and worst kayakers' performance, the best showed greater forces magnitudes and synchronization of the peak forces. Analyses of the force–time curves, including not only the forces applied by the kayaker on the paddle but also the ones applied on the foot rest and strap, should be considered relevant in terms of technique analyses.

Keywords: kayaking; kinetics; lower limbs; load cells; foot strap; foot rest; force measurements

Citation: Bonito, P.; Sousa, M.; Ferreira, F.J.; Justo, J.F.; Gomes, B.B. Magnitude and Shape of the Forces Applied on the Foot Rest and Paddle by Elite Kayakers. *Sensors* **2022**, *22*, 1612. https://doi.org/10.3390/s22041612

Academic Editor: Vesa Linnamo

Received: 2 January 2022
Accepted: 15 February 2022
Published: 18 February 2022

1. Introduction

In sprint kayaking, the athlete produces force to move the kayak forward, and these forces are transmitted to the water by the paddle blade and to the kayak by the foot rest, foot strap, and seat [1]. The introduction of the "wing" blade originated changes in the paddling technique with a tendency for a lateral paddle path during the water phase of the stroke, which increased the contribution of the trunk rotation and a great lower limbs movement to force development and performance [1]. The greater lower limb action and rotation of the pelvic girdle have been associated with the performance of more experienced/high-level kayakers [2–4]. When comparing performances involving restricted and unrestricted pelvic girdle and lower limbs movement on a kayak ergometer, unrestricted movements were associated with better blade velocities (+0.2 to 0.4 $m \cdot s^{-1}$), higher paddling impulse (+3.5 N·s), significant increases in strength throughout the paddle cycle, decrease in mechanical work (−4%), better propulsion (+6%), and consequently an increase in the performance (+6%) [5]. These results were confirmed in an on-water kayaking study, also comparing restricted and unrestricted, and concluded that the restriction of lower limbs leads to a significant reduction of the kayak velocity (−16%) and a decrease of the bilateral mean stroke force (−21%), however, in terms of the time pattern of the occurrence of the peak force on the foot rest and foot strap, there were no significant changes [1]. Futhermore, the impulse of the push forces on the foot rest over 10 s showed the highest correlation to maximum kayak velocity [6].

A small number of studies have analyzed kayaking kinetics and its relationship with performance on a kayak ergometer [2] and on-water situation [7,8] and were more centered on the forces applied on the paddle. In these studies, the best performances were associated with paddle force–time curves where the kayakers reached peak force quickly and tended to maintain a force near peak values until the exit of the paddle from the water. Gomes and colleagues [7] reported that the mean kayak velocity had a higher correlation with mean paddle force compared with peak force, therefore, when the area under the force–time curve increased and the stroke frequency was maintained, the speed of the kayak increased. They also reported that the elite paddlers' performance was characterized by shorter water phase duration, higher handgrip loading, higher blade force, and impulse compared to a recreational paddler [7,8].

With regard to the study of the forces applied on the paddle, but also on the foot rest and strap, Tornberg and colleagues [9] analyzed these forces in a pilot study and reported that high-level kayakers produced the same forces on the paddle compared to the low level ones, but in relation to the forces applied on the foot rest, the high level ones produced 3 to 26 times more force. A subsequent study [10], which also intended to analyze these forces but on water, showed that the forces applied on the foot rest began to be applied before the paddle forces on each stroke. However, this study could not report in detail the magnitude of the forces applied on the foot rest or strap, being only valid for the timing of the different movements [10].

A few studies [1,6,9] have been dedicated to the analyses of the forces applied on the foot rest and foot strap, and none have analyzed the magnitude and shape of the force–time curves of compression (push foot rest forces), tension (pull foot strap forces), and paddle, as well as the synchronization between these forces in elite kayakers. Therefore, the goals of the present study were to describe the magnitude and shape of the force–time curves of compression and tension applied on the foot rest and foot strap, respectively, by studying elite kayakers and their performance on a kayak ergometer, and simultaneously, to analyze the synchronization and timing of these forces with the paddle force–time curves.

2. Analysis of Interaction between the Kayaker, Paddle, Kayak and Water

To improve the kayaker's performance, it is essential to understand the mechanics of interaction forces between the kayaker, paddle, kayak, and water. To assess the efficiency of the paddling movement, it is vital to quantify all the actions (forces and moments) that the kayaker exerts during this process. Figure 1 shows all the actions on these three bodies, where $\vec{W_T}$ is the total weight of the three bodies; $\vec{P}$, paddle propulsion force; $\vec{D_w}$, water drag force; $\vec{D_a}$, air drag force; $\vec{B}$, buoyant force; $\vec{M_B^z}$, moment caused by water distributed on the hull. Considering their small magnitude, $\vec{D_a}$ and $\vec{M_B^z}$ were neglected.

This study focused on the following forces developed by the kayaker: $\vec{P}$, paddle propulsion force; and foot rest forces, $\vec{F_L}$ and $\vec{F_R}$. The interaction forces between kayaker and kayak in the Oxy plane are shown in Figure 2. The kayak seat receives the main weight of the kayaker, whereas the foot rest and foot strap receive the main forces developed by the kayaker legs/feet. The foot rest force applied by one foot, combined with the foot strap force applied by the other foot, creates a moment along the *z*-axis that makes it possible for the kayaker to rotate its pelvic girdle for paddling left and right and counterbalances the moment created by the paddle propulsion force, $\vec{P}$ (Figure 2).

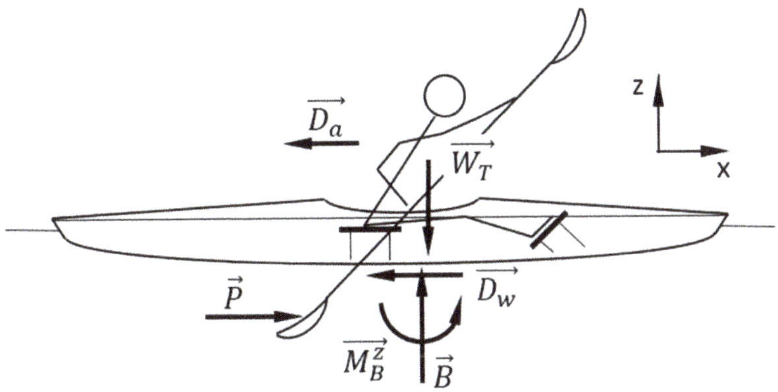

Figure 1. Kayaker/kayak/paddle free body diagram.

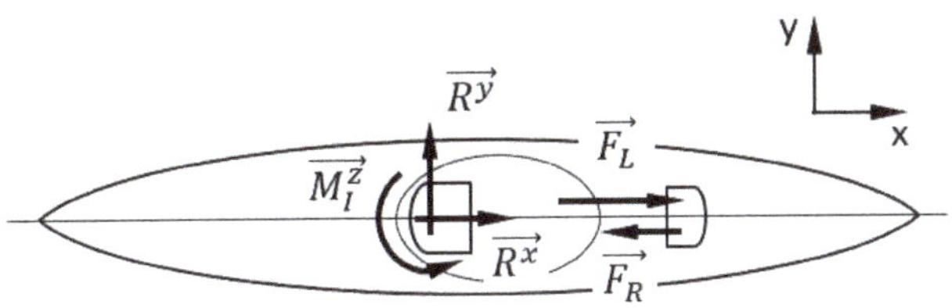

Figure 2. Kayaker/kayak interaction diagram.

3. Materials and Methods

3.1. Instrumentation and Devices

The foot rest/strap was custom-designed and built at the ISEP/IPP, School of Engineering. This foot rest/strap (Figure 3) has a structure similar to the shape of the original foot rest and is divided into two parts, each one having four double strain gauge load cells (Figure 4), allowing measurement of the left and right forces independently. These load cells work in both directions, measuring push and pull forces (compression and tension forces, respectively). The position of the four load cells on each side was chosen so they would work similarly to the supports of a scale. The load cells are connected in a full Wheatstone bridge with 8 active strain gauges, resulting in an instrumented foot rest that is insensible to the location of the applied force.

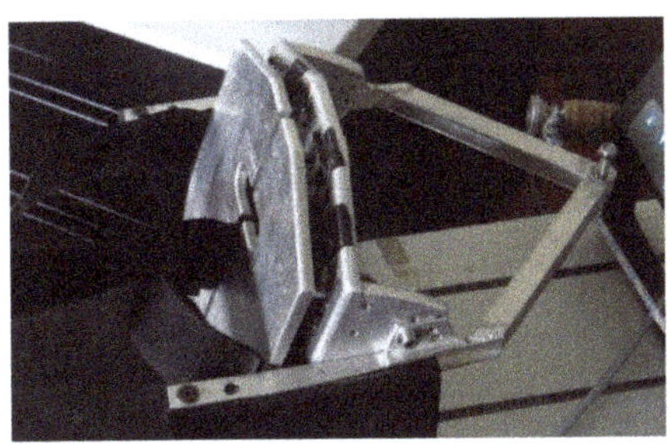

Figure 3. Custom-built foot rest, that measure push and pull forces over the foot rest and foot strap, respectively.

(A) (B)

Figure 4. (**A**) View of the foot rest's interior with four load cells on each side, (**B**) the detail of one load cell.

Each side of the foot rest was calibrated in the laboratory using calibrated weights and a Micro-Measurements VISHAY P3 signal conditioner (Figure 5).

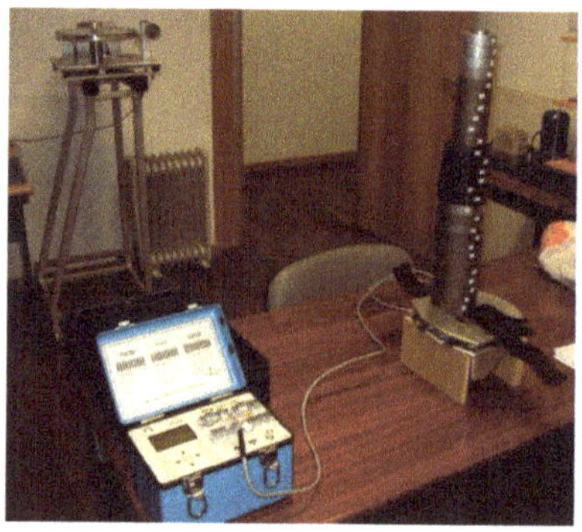

Figure 5. Foot rest calibration procedure.

The ergometer paddle drive ropes were instrumented with load cells (Figure 6) custom-designed and built for this application. These load cells were designed to work in tension, their mechanical behavior was assessed using finite element software. They were manufactured from an aluminum alloy 5083 piece in a CNC machine at ISEP/IPP. Each one has 4 strain gauges in a full Wheatstone bridge configuration. Both were calibrated at ISEP/IPP laboratory using calibrated weights and the signal conditioner VISHAY P3.

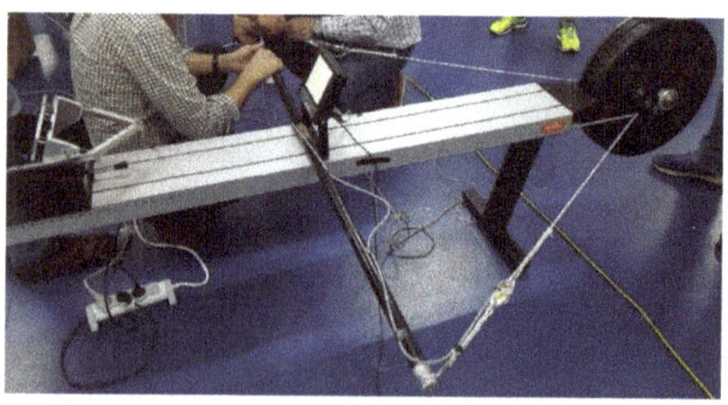

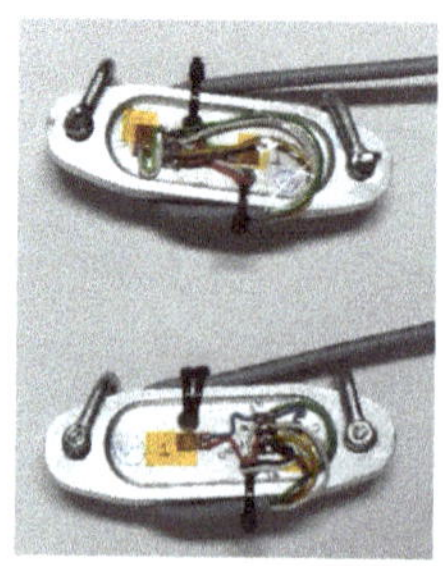

(A) (B)

Figure 6. (**A**) Kayak ergometer with instrumented foot rest and paddle ropes, (**B**) custom design load cells built for measuring kayak ergometer paddle ropes forces.

3.2. Participants

The study was conducted with thirteen elite male kayakers (5 senior and 8 under-23 athletes), age 23.3 ± 4.9 years, height 179.0 ± 5.7 cm, and a body mass of 79.70 ± 5.70 kg. The senior kayakers had international participations, including World Championships and Olympic Games. The under-23 kayakers had participated in International junior and under-23 competitions in the last two years. The participants were selected considering the performance times in the 500 m of the Sprint National Cup and participation in World Cups and World Championship. Subjects were fully informed of the nature of the investigation and provided written informed consent before data acquisition. The study was approved by the Institutional Review Board performed according to the Declaration of Helsinki.

3.3. Procedures

To observe and describe the magnitude and shape of the forces applied on the foot rest, foot strap, and paddle, a kayak ergometer (Dansprint Pro, Denmark) [11] was instrumented and data from the forces applied on the foot rest and strap, compression and tension, were synchronized with data of the tension on the paddle drive ropes. Kayak ergometer has been considered a valid tool for physiological testing [12]. In terms of biomechanical demands, these probably do not fully replicate [13], however, it has been shown that there are no significant differences in terms of paddle force variables—peak force, time to reach peak force, and impulse [14]. The kayakers that participated in the present study were familiar with the ergometer, considering that they are often used for training, mainly in the winter season, and that their training control tests are usually performed on a kayak ergometer. The drag of the ergometer was adjusted considering the kayaker's body mass as recommended by the Dansprint to simulate on-water paddling drag.

A 2-min test race pace was performed on the kayak ergometer simulating 500 m sprint race performance. The warm-up consisted of a high-intensity interval exercise for kayak ergometer adapted from Bishop and colleagues [15]. At the end of the warm-up, the athletes rested passively for 5 min and then performed the test. After the 2-min test they performed a 10-min active recovery on a static bike at 60 rpm.

3.4. Kinetic Data Collection

Data processing and storage was performed using a PC and a signal conditioner system (HBM® Spider8, Darmstadt, Germany). Data was recorded at 100 Hz sample rate. The system synchronously recorded data from all channels (left and right foot rest, left and right paddle ropes), as the different channels were connected to the signal conditioner using the same time base for all channels.

3.5. Data Analysis

The data was processed by MATLAB R2016a (The MathWorks Inc., Natick, MA, USA) and analyzed using an in-house developed routine. The routine automatically detected the strokes by analyzing both the force data from the load cells on the paddle rope (left and right) synchronized with the foot rest/strap load cells. Paddle (P) and foot rest/strap forces (expressed in newtons) were analyzed for each side (l, left and r, right). On the foot rest/strap the compression (C) forces represented positive values of the curve in the direction of kayak travel considering the pull forces applied to the foot rest, and the tension (T) forces represented negative values of the curve, in the opposite direction by pulling the foot strap. The MATLAB routine identified the compression stroke onset when the force applied on the foot rest/strap crossed from negative to positive, and its end, when again the force went from positive to negative. The opposite was used for the tension forces, so that the onset occurred when the force applied on the foot rest/strap went from positive to negative and its end when the values went positive again. For the cable rope force analysis, the onset and end of the paddle stroke in the left and right sides was defined when, for two consecutive values, the data was above or below 10 N, respectively.

In each of the force curves (Figure 7), and considering all the paddle strokes, the peak force (PF), mean force (MF), and impulse (I) were computed. The PF represented the maximum value reached, the MF the average value of the forces and impulse the area below the force curve (force–time integral). The foot rest and strap variables that were considered for analysis were:

(1) left and right peak force compression (CPF) over the foot rest, and left and right peak force tension (TPF) over the foot strap;
(2) left and right compression mean force (CMF) over the foot rest, and left and right tension mean force (TMF) over the foot strap; and
(3) left and right compression impulse (CI) over the foot rest, and left and right tension impulse (TI) over the foot strap.

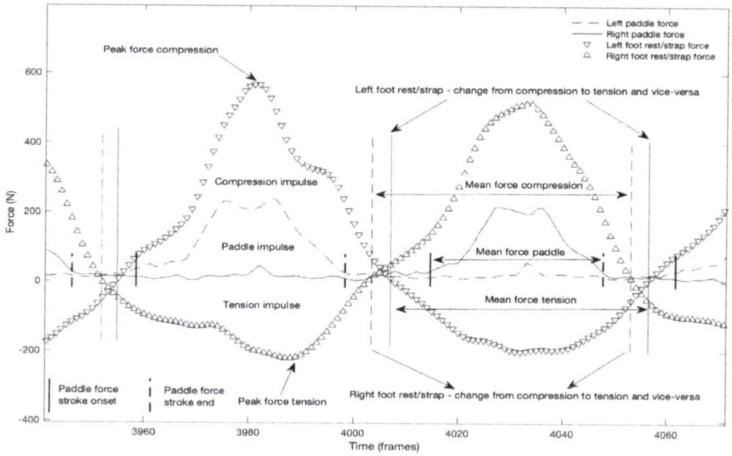

Figure 7. Example of two complete paddle strokes (left and right) presenting the paddle, foot rest, and foot strap force curves synchronized, and the main kinetic variables considered for analysis.

The previous variables (PF, MF, and I) were also studied for the paddle force curves: (1) left and right peak paddle force; (2) left and right mean paddle force; and (3) left and right paddle impulse.

The mean force/peak force ratio (MF/PF R) was computed for compression and tension curves, as well as for paddle force curves, and reflects the force profile, where a rectangular shape is represented by 100% and closer to 50% is a tendency for a triangular shape [16].

Time variables were also computed for each stroke: (1) the duration in time of the compression and tension forces, as well as paddle force application in each side; and (2) the time to peak force of the compression and tension forces applied to the foot rest/strap and on the paddle.

The performances with the higher and lower mean power reached on the kayak ergometer (given by the kayak ergometer software) during the 2-min test were analyzed in detail. When comparing the best with worst performance, the variables mentioned above were analyzed in absolute and also normalized considering the kayakers' weight.

3.6. Statistics

All statistical analysis was conducted using SPSS v25.0 (SPSS Inc., Chicago, IL, USA). Descriptive statistics included mean and standard deviation. The data were checked for distribution normality and homoscedasticity with shapiro–Wilks and Levene tests, respectively. T-student for paired samples was used to compare differences within each variable (foot rest/strap and paddle) between left and right sides. Pearson correlation test was computed for the analysis of the correlations between these variables and also to correlate left and right sides forces. A correlation matrix was designed on MATLAB R2016a (The MathWorks Inc., Natick, MA, USA). The confidence interval was set at 95%. For the analysis of the differences between the best and worst performances, the difference in percentage for each variable was computed—Dif. % = ((BF − WP)/((BF + WP)/2) × 100), where BF is the result in each variable of the kayaker's best performance and WP is the kayaker with the worst performance.

4. Results

4.1. Analysis of the Magnitude and Shape of the Force Applied on the Foot Rest/Strap and Paddle

The mean power generated by the kayaker on the kayak ergometer was set as the performance indicator (Table 1). Table 2 presents the magnitude values of the forces applied on the foot rest and strap, time variables, and the profile of the force curves, considering the ratio mean force/peak force. No differences were observed, when comparing the mean values of the forces applied on the left and right foot rest and foot strap. The magnitude of the peak forces of compression over the foot rest more than doubled the tension peak forces applied over the foot strap; the same was observed for impulse. The duration of force application was higher for compression than tension. Since left pull (compression) occurs simultaneously with right push (tension) over the foot rest/strap, and vice-versa, the data showed a similar mean difference in duration (0.07 s). The force profile (MF/PF R) for the compression forces tended towards a triangular shape (≈53%) and for tension forces tended more towards a rectangular shape (≈63%).

Table 1. Mean power (W) and mean stroke rate (stroke per minute) of each athlete and overall mean in the 2-min kayak ergometer performance test.

Kayaker	Mean Power (W)	Mean Stroke Rate (spm)
1	286	123
2	267	115
3	341	119
4	322	118
5	292	126
6	253	126
7	307	112
8	307	123
9	294	127
10	292	122
11	258	120
12	264	123
13	260	125
Mean ± SD	288 ± 26.9	121 ± 4.5

W, watt; spm, strokes per minute; SD, standard deviation.

Table 2. Variables related to the forces applied on foot rest (compression) and strap (tension) for left and right sides (mean ± SD). Positive values represent compression, and negative values represent tension forces.

Foot Rest/Strap Forces	Left	Right
	Mean ± SD	Mean ± SD
CPF (N)	543.27 ± 85.93	524.39 ± 88.36
TPF (N)	−240.09 ± 74.92	-231.05 ± 52.01
CMF (N)	283.04 ± 50.76	279.09 ± 64.13
TMF (N)	−152.72 ± 49.47	−147.74 ± 35.36
CI (N·s)	143.62 ± 32.66	148.21 ± 31.89
TI (N·s)	-72.77 ± 27.01	−65.94 ± 22.17
CFD (s)	0.52 ± 0.07	0.55 ± 0.05
TFD (s)	0.48 ± 0.06	0.45 ± 0.06
T.CPF (s)	0.32 ± 0.06	0.31 ± 0.05
T.TPF (s)	0.24 ± 0.08	0.21 ± 0.03
MF/PF RC (%)	52.31 ± 5.84	53.15 ± 5.68
MF/PF RT (%)	63.31 ± 4.29	63.92 ± 2.84

CPF, compression peak force; TPF, tension peak force; CMF, compression mean force; TMF, tension mean force; CI, compression impulse; TI, tension impulse; CFD, compression force duration; TFD, tension force duration; T.CPF, time to reach compression peak force; T.TPF, time to reach tension peak force; MF/PF RC, compression mean force/peak force ratio; MF/PF RT, tension mean force/peak force ratio; SD, standard deviation.

Data from the analyses of the forces applied on the paddle are presented in Table 3. There were no statistical differences between the magnitude of the forces applied on the left and right paddle sides ($p > 0.05$).

Table 3. Variables related to the forces applied on the paddle on the left and right sides.

Paddle Forces	Left	Right
	Mean ± SD	Mean ± SD
PPF (N)	236.37 ± 19.32	243.92 ± 28.89
PMF (N)	88.53 ± 6.87	88.68 ± 6.58
PI (N·s)	43.22 ± 4.76	42.23 ± 3.32
PFD (s)	0.49 ± 0.03	0.49 ± 0.04
T.PPF (s)	0.21 ± 0.02	0.21 ± 0.03
MF/PF RP (%)	38.82 ± 2.59	38.26 ± 5.27

PPF, paddle peak force; PMF, paddle mean force; PI, paddle impulse; PFD, paddle force duration; T.PPF, time to reach paddle peak force; MF/PF RP, paddle mean force/peak force ratio.

The performance (mean power on the kayak ergometer) presented a positive correlation with the paddle mean force (r = 0.792, $p < 0.001$) and paddle impulse (r = 0.701, $p < 0.001$), respectively (Figure 8). The stroke rate presented a negative correlation with paddle impulse (r = −0.530, $p < 0.01$). Also, a correlation was found between the variables in analyses of the paddle stroke, when analysing left and right sides (peak force—r = 0.786, $p < 0.01$; mean force—r = 0.670, $p < 0.05$; impulse—r = 0.648, $p < 0.05$).

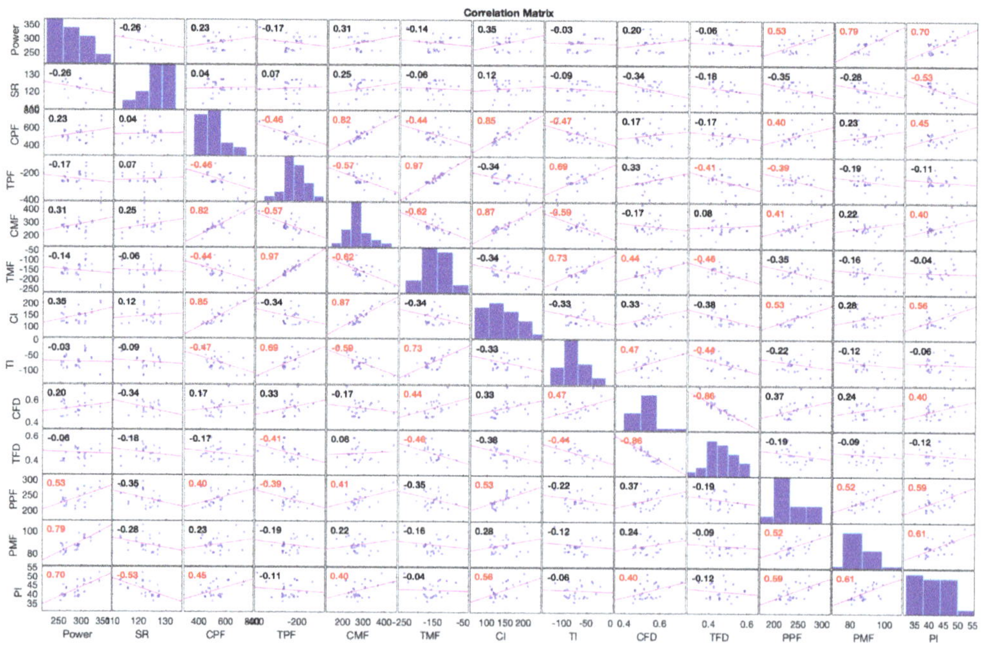

Figure 8. The correlation matrix with correlation coefficients between power and force variables of the paddle, foot rest and foot strap, combining left and right of 13 kayakers. Red correlation coefficients indicate strong correlation, the tendency line shows if its positive or negative. SR—stroke rate; CPF—compression peak force; TPF—tension peak force; CMF—compression mean force; TMF—Tension mean force; CI—compression impulse; TI—Tension impulse; CFD—compression force duration; TFD—tension force duration; PPF—paddle peak force; PMF—paddle mean force; PI—paddle impulse.

The compression peak force correlated negatively with the tension peak force applied on the opposite side of the foot rest/strap (r = −0.461, $p < 0.05$) and the same was observed for the analysis of the mean force values, where compression mean force correlated negatively with the tension mean force applied on the opposite side of the foot rest/strap (r = −0.613, $p < 0.001$). Left and right compression mean forces were also correlated (r = 0.874, $p < 0.01$). When variables related to the forces applied to the foot rest and strap were correlated with paddle forces, paddle peak force showed a positive correlation with the compression peak force (r = 0.400, $p < 0.05$) and with the compression impulse (r = 0.534, $p < 0.01$). Compression impulse was correlated with the left paddle impulse (r = 0.672, $p < 0.01$).

4.2. Comparative Analysis of the Best and Worst Performances

Mean power during the 2-min test was used as parameter to determine the best and worst performances (the difference in mean power between the two athletes was 88 W). When analyzing the mean force–time curves and results of the kayakers with the best (Figure 9) and the worst (Figure 10) performances, the results suggest differences, both in terms of magnitude and profile.

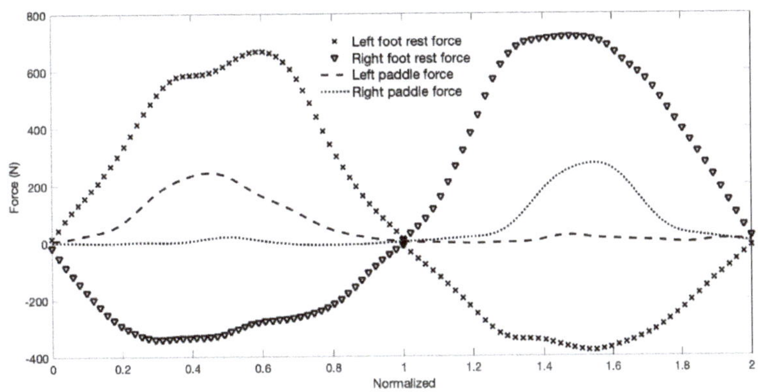

Figure 9. Mean force–time curves of the forces applied on the foot rest/strap and paddle of the kayaker with the best performance of the sample group. Data was time normalized and represents N = 238 strokes (119 complete stroke cycles).

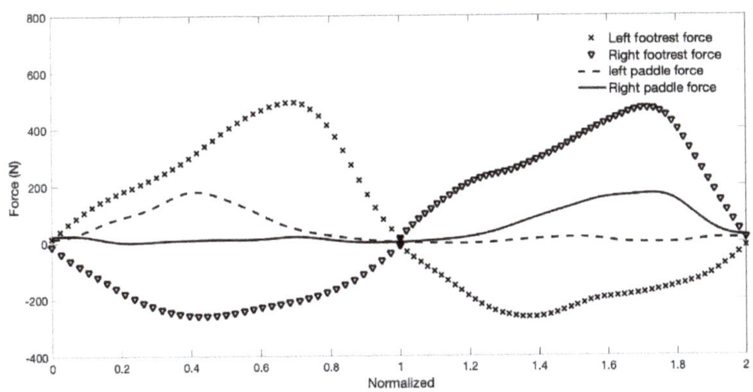

Figure 10. Mean force–time curves of the forces applied on the foot rest/strap and paddle of the kayaker with the worst performance of the sample group. Data was time normalized and represents N = 252 strokes (126 complete stroke cycles).

The visual comparison of Figures 9 and 10 suggests different ways and timings of force development. The best performance presents a tendency for curves with higher values, and in middle of the curve, a tendency for a plateau; the analyses also shows more pronounced slopes from and to zero force. The kayaker with the best performance reaches the peak forces on the paddle and foot rest from the same side and foot strap of the opposite side almost simultaneously, which is no longer observed in Figure 10, which presents the data of the athlete with the worst performance.

Table 4 presents the magnitude of the forces applied by the two kayakers, best and worst performances. Results show higher absolute values for the kayaker with the best performance, for each of the variables analyzed in terms of foot rest, foot strap, and paddle forces (peak and mean force, and impulse) when compared with the results from the kayaker with the worst performance. When the variables area presented was normalized to paddlers weight, the difference in % were lower and mainly in foot rest/strap force variables.

Table 4. The magnitude of the forces applied on the foot rest/foot strap and on the paddle by the two kayakers, with the best and worst performances. Absolute values and relative values to the kayakers' weight, both presenting the differences in % between kayaker. Best performance n = 238 strokes and worst performance n = 252 strokes.

Foot Rest/ Strap Forces	Absolute Values		Dif. %	Foot Rest/ Strap Forces	Relative to Weight		Dif. %
	Best	Worst			Best	Worst	
lCPF (N)	676.78	501.56	30	lCPF (N·kg^{-1})	7.52	6.69	12
rCPF (N)	736.43	487.66	41	rCPF (N·kg^{-1})	8.18	6.50	23
lTPF (N)	−388.11	-267.14	37	lTPF (N·kg^{-1})	−4.31	−3.56	19
rTPF (N)	−345.16	-261.23	28	rTPF (N·kg^{-1})	−3.84	−3.48	10
lCMF (N)	387.57	272.47	35	lCMF (N·kg^{-1})	4.31	3.63	17
rCMF (N)	441.47	274.4	47	rCMF (N·kg^{-1})	4.91	3.66	29
lTMF (N)	-243.44	-167.33	37	lTMF (N·kg^{-1})	−2.70	−2.23	19
rTMF (N)	-230.99	−171.5	30	rTMF (N·kg^{-1})	−2.57	−2.29	12
lCI (N·s)	201.96	118.95	52	lCI (N·s·kg^{-1})	2.24	1.59	34
rCI (N·s)	209.02	154.26	30	rCI (N·s·kg^{-1})	2.32	2.06	12
lTI (N·s)	−114.46	-85.44	29	lTI (N·s·kg^{-1})	−1.27	−1.14	11
rTI (N·s)	−119.53	-66.09	58	rTI (N·s·kg^{-1})	−1.33	−0.88	41
Paddle Paddle Forces	**Absolute Values**		**Dif. %**	**Paddle**	**Relative to Weight**		**Dif. %**
	Best	**Worst**			**Best**	**Worst**	
lPPF (N)	273.4	225.3	19	lPPF (N·kg^{-1})	3.0	3.0	0
rPPF (N)	293.6	263.2	11	rPPF (N·kg^{-1})	3.3	3.5	−6
lPMF (N)	105.3	80.9	26	lPMF (N·kg^{-1})	1.2	1.1	9
rPMF (N)	99.1	88.9	11	rPMF (N·kg^{-1})	1.1	1.2	−9
lPI (N·s)	52.3	36.5	36	lPI (N·s·kg^{-1})	0.6	0.5	18
rPI (N·s)	47.0	39.4	18	rPI (N·s·kg^{-1})	0.5	0.5	0

Foot rest forces: lCPF, left compression peak force; rCPF, right compression peak force; lTPF, left tension peak force; rTPF, right tension peak force; lCMF, left compression mean force; rCMF, right compression mean force; lTMF, left tension mean force; rTMF, right tension mean force; lCI, left compression impulse; rCI, right compression impulse; lTI, left tension impulse; rTI, right tension impulse; paddle forces: lPPF, left paddle peak force; rPPF, right paddle peak force; lPMF, left paddle mean force; rPMF, right paddle mean force; lPI, left paddle impulse; rPI, right paddle impulse: Dif.%, difference in % between kayakers.

5. Discussion

This is the first study that combines paddle, foot rest and strap kinetic analysis in a group of elite kayakers. Previous studies have focused on the paddle or foot rest forces independently, however this issue has never been addressed in relation to performance [1,5,17]. Therefore, the main goal of the present study was to characterize the profile of the curves and magnitude of the forces exerted by the lower limbs of elite kayakers during 2 min on a kayak ergometer, simulating sprint race performance, synchronized with the forces applied on the paddle. Results show that in each stroke the forces applied on the foot rest and foot strap begun to be applied before the ones applied on the paddle shaft, as previously reported in a study testing only two kayakers (one male and one female) and performed on water [10].

Analyzing the magnitude of the forces, in terms of peak forces, the ones applied on foot rest (compression) doubled the ones applied on the paddle. Already, the forces on the foot strap (tension) were similar in magnitude to paddle forces. When analyzed in terms of mean values, the compression forces were more than three times and the tension forces almost two times, the forces applied on the paddle. The differences were higher when analyzed in terms of the mean values, compared with peak, due to the fact that in terms of curve shape, the forces applied on the foot rest and foot strap had a much higher tendency for a rectangular profile, instead of being closer to the triangular profile of the paddle forces. Another aspect to note was that there were no significant differences between left and right-side forces, for both the foot rest and foot strap as well as paddle, prevailing the idea of non-significant asymmetry in elite kayakers [17,18].

No direct and strong correlation was observed between the foot rest forces (magnitude and time-dependent variables) and the performance (power) as previously reported [19], except the correlation between the left compression force duration and the performance ($r = 0.560$, $p < 0.05$). If a longer compression force duration would be associated with a greater range of motion of the knee, results are in line with a previous study [3], that suggested that greater knee amplitude are associated with elite kayakers' performance. Nilsson and colleagues [1] observed that kayak speed decreased by 16% with the restriction of movement of the lower limbs and also reported a high correlation between pushing forces and kayak velocity, which suggests that the force applied by the lower limbs considerably influences the kayaking performance [6]. Previously, Begon and colleagues [5], based on computer modeling to access the lower limb contribution to kayak performance, suggested that the legs produce 6% of the total propulsion on a sliding kayak ergometer. The correlation analysis of the compression and tension forces applied simultaneously (one side foot rest and the foot strap of the opposite side) led to the suggestion that the higher the compression force, the greater the tension force on the opposite side. Although this is the first study to show a positive linear correlation between the magnitude of the forces applied on the foot rest and foot strap, it is not possible, however, to define a cause–effect relation.

The analyses of the forces applied on the paddle showed that performance (mean power in kayak ergometer) correlated with paddle mean force ($r = 0.792$, $p < 0.001$) and paddle impulse ($r = 0.701$, $p < 0.001$). Was also observed a high correlation between these variables and kayak velocity [7], which is in line with the results of the present study. A significant correlation between performance and peak paddle force (lower correlation) was also presented, which is in line with the results of other paddle force profile studies in on-water situations [7]. Maintaining force values near the peak force throughout the water phase is more important than the peak power itself, thus generating a more effective paddle stroke and consequently propelling the kayak even more [2].

The paddle mean force/peak force ratio did not correlate with ergometer performance, however, previous studies concluded that the best performances were associated with higher values of this ratio [7,19]. Although the performance was correlated with the mean and impulse paddle forces and not with peak force, it suggests a tendency for a rectangular profile. However, this did not occur in the present study, probably as suggested by Begon and colleagues [19], due to the low forces values recorded before and after the supposed water entry/exit of the blade (Figure 11). These low values of tension on the kayak rope probably have interfered with the mean paddle force values and consequently with the mean force/peak force ratio. The low value, before and after the presumed delimitation of paddle stroke, was due to the recoil spring used in the ergometer, which continuously creates tension for collecting the cable. The use of a kayak ergometer instead of on-water performance was a limitation of the present study, however, it allowed to use an instrumented foot rest with a high number of load cells, although heavy, that collected compression and tension forces, which reflected the kayak force application by elite kayakers. Future studies should focus on the analysis in an on-water situation, collecting not only paddle and foot rest and strap forces, but also adding seat forces that is the other point of contact with the kayak.

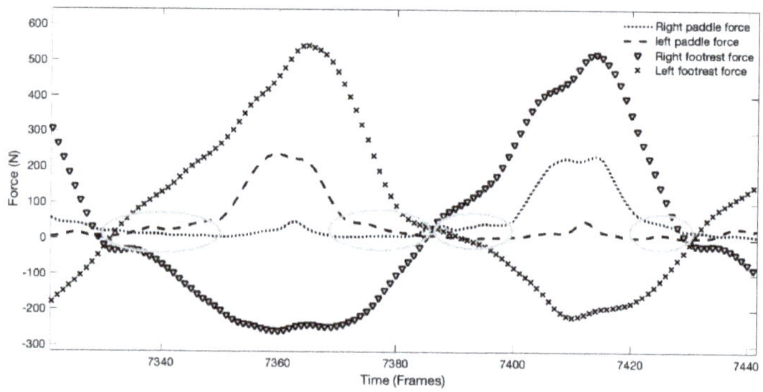

Figure 11. Raw data of the force–time curves of the forces applied on the foot rest/foot strap and paddle, showing the values close to zero before entry and after exit in water no the paddle force curve, identified with grey ellipses.

The increase in paddle peak force was related to the increase of the compression peak force and impulse on the side of the stroke. Kayakers with higher paddle impulses also had higher compression impulses on the side of the stroke, suggesting that there is a dependence between the magnitude of the forces applied by the upper and lower limbs, on paddle and foot rest, respectively, as already had been previously suggested in a study in kayak ergometer [5] and lately confirmed in on-water situation [1]. Considering the correlations observed between the foot rest and strap forces with the forces applied on the paddle, and considering that these last were also correlated with performance, it suggests that, to a certain extent, the forces applied on the foot rest and strap are indirectly related to the performance. This is the first study to focus on the relation between the forces applied on paddle, foot rest, and strap, a to analyze it with respect to performance.

The analyses of the force–time curve shape showed that the forces of compression and tension (on opposite sides) began to be applied before forces being applied on the paddle, as also observed in a pilot study [9]. Therefore, each paddle stroke starts with lower limbs force development. Both forces applied on foot rest and strap also ended after the paddle force end, continuing until a certain moment of the aerial phase where both legs change from compression to tension and vice-versa, initiating the paddle stroke on the opposite side.

The compared analyses of the forces applied on the paddle, foot rest and foot strap, considering the best and worst performances (Figures 9 and 10), showed differences in magnitude and shape of the force–time curves and consequently differences in performance. When considering the magnitude of the forces, results suggest that the greater magnitude can be associated with the best performance for both absolute and relative weight analyses of the forces. Another very interesting aspect was the timing in which the peak forces of compression, tension (on opposite sides) and on the paddle occurs (Figures 9 and 10). For the kayaker with the best performance, the peak timing of the three forces tended to be simultaneous, unlike the kayaker with the worst performance. It was also observed in a previous study that international level kayakers showed closer peak synchronization compared to national elite kayakers [10]. However, it would be of interest to confirm the present results, since they suggest that better performances are dependent on the ability to develop higher forces, but also to be able to synchronize their peaks.

Performance in kayaking depends on the magnitude and the way forces are applied by the kayaker on the paddle and points of contact with the kayakers, but also the way this reduces drag forces as much as possible. Therefore, a higher magnitude of forces applied is not an exclusive condition that allows defining which will be the best performance, since it will have to be balanced with the drag forces that are being generated. In this way, kayakers with similar performances can have considerable differences in the magnitude of the forces applied, mainly to the foot rest and strap.

The comparison of the results of the kayakers with the best and worst performances showed that the one with the worst performance tended to take longer time to reach the peak force (low slope) of both left and right compressions and peak paddle force, which have been suggested as factors that limit the performance [7]. Already, the best performance showed a force curve shape where the peak forces were reached quickly and tended to remain for a period near the peak (plateau), which have been mentioned as a determinant of kayaking efficiency [1]. Also, the results in terms of ratios (mean force/peak force), for the best performance kayaker, showed a tendency for a more rectangular shape of the force–time curves [16].

The study presents unique data about the forces applied on the foot rest, strap, and paddle, correlating them with performance in kayak ergometer. It was shown that the forces applied on the foot rest and strap in each paddle stroke began to be applied earlier than the forces on the paddle and ended later, continuing till a certain moment of the aerial phase where both legs change from compression to tension and vice-versa, to initiate the paddle stroke on the opposite side. It was also observed that the higher the magnitude of the paddle forces, the better the performance. The paddle forces were found to be positively related to the magnitude of the forces applied on the foot rest, suggesting an indirect relation between the forces applied on the foot rest and the kayaking performance. When analyzing the shape of the force–time curves, the results suggest that best performances had a higher tendency for a rectangular shape, besides a tendency for a synchronized occurrence of the peak forces applied on the foot rest and paddle, on the same side, and on the foot-strap on the opposite side.

6. Conclusions

The analyses of the force–time curve, magnitude and shape, should be a prime objective in terms of technique analyses, not only the forces applied on the paddle but together with the forces applied by the kayaker on the foot rest and strap. The difference is that the force applied on the foot rest and strap is maintained continuously along a stroke cycle, changing each leg from compression to tension rapidly, while paddle force is applied intermittently, with periods of no force being applied. Scientists and coaches, regarding kayak technique and performance, should also focus on lower body performance, considering that the forces measured on the foot-rest doubled the ones applied on the paddle, besides the fact that the best performances had a higher magnitude of forces not only applied on paddle but also on the foot rest/strap. This study, as well as similar studies, have the potential to improve knowledge on the propulsion forces in kayaking, and consequently, help coaches and athletes improve sport performance. The deep knowledge about how elite kayakers apply forces (both magnitude and shape) and how the application of these forces are synchronized is also of interest. Understanding what the best kayakers are able to do, above the ones that are already in the elite group, can guide investigators to the precise aspects that influence performance.

Author Contributions: Conceptualization, P.B., F.J.F. and B.B.G.; methodology, P.B., F.J.F., J.F.J. and B.B.G.; software, F.J.F., J.F.J. and B.B.G.; validation, M.S., F.J.F. and J.F.J.; formal analysis, P.B., F.J.F., J.F.J. and B.B.G.; investigation, P.B., F.J.F., J.F.J. and B.B.G.; resources, P.B., F.J.F., J.F.J. and B.B.G.; data curation, P.B. and B.B.G.; writing—original draft preparation, P.B. and B.B.G.; writing—review and editing, P.B., M.S., F.J.F., J.F.J. and B.B.G.; supervision, F.J.F. and B.B.G. All authors have read and agreed to the published version of the manuscript.

Funding: This research received no external funding.

Institutional Review Board Statement: The study was conducted in accordance with the Declaration of Helsinki and approved by the Institutional Review Board (08/11/2017 CC 265/2017) of the Faculty of Sports Sciences and Physical Education of the University of Coimbra.

Informed Consent Statement: Informed consent was obtained from all participants involved in the study.

Data Availability Statement: The data presented in this study are available on request from the author.

Acknowledgments: The authors are grateful for the kind collaboration of the kayakers who participated in the study. We thank the collaboration, on data collection, of Élio Henriques, Gonçalo Neves, José Galvão, Rui António Fernandes, and Tiago Viegas. We also thank the support of the Portuguese Canoe Federation and National coaches.

Conflicts of Interest: The authors declare no conflict of interest.

References

1. Nilsson, J.E.; Rosdahl, H.G. Contribution of Leg-Muscle Forces to Paddle Force and Kayak Speed During Maximal-Effort Flat-Water Paddling. *Int. J. Sports Physiol. Perform.* **2016**, *11*, 22–27. [CrossRef] [PubMed]
2. Michael, J.S.; Smith, R.; Rooney, K.B. Determinants of kayak paddling performance. *Sports Biomech.* **2009**, *8*, 167–179. [CrossRef] [PubMed]
3. Limonta, E.; Squadrone, R.; Rodano, R.; Marzegan, A.; Veicsteinas, A.; Merati, G.; Sacchi, M. Tridimensional kinematic analysis on a kayaking simulator: Key factors to successful performance. *Sport Sci. Health* **2010**, *6*, 27–34. [CrossRef]
4. Lee, C.H.; Nam, K.J. Analysis of the kayak forward stroke according to skill level and knee flexion angle. *Int. J. Bio-Sci. Bio-Technol.* **2012**, *4*, 41–48.
5. Begon, M.; Colloud, F.; Sardain, P. Lower limb contribution in kayak performance: Modelling, simulation and analysis. *Multibody Syst. Dyn.* **2010**, *23*, 387–400. [CrossRef]
6. Klitgaard, K.K.; Rosdahl, H.; Brund, R.B.K.; Hansen, J.; Zee, M. Characterization of Leg Push Forces and Their Relationship to Velocity in On-Water Sprint Kayaking. *Sensors* **2021**, *21*, 6790. [CrossRef] [PubMed]
7. Gomes, B.; Ramos, N.V.; Conceição, F.A.V.; Sanders, R.H.; Vaz, M.A.P.; Vilas-boas, J.P. Paddling Force Profiles at Different Stroke Rates in Elite Sprint Kayaking. *J. Appl. Biomech.* **2015**, *31*, 258–263. [CrossRef] [PubMed]
8. Niu, L.; King, P.W.; Tay, C.S.; Lin, Y.; Wu, B.; Ding, Z.; Chan, C.C. Evaluating On-Water Kayak Paddling Performance Using Optical Fiber Technology. *EEE Sens. J.* **2019**, *19*, 11918–11925. [CrossRef]
9. Tornberg, A.B.; Hakansson, P.; Svensson, I.; Wollmer, P. Forces applied at the footrest during ergometer kayaking among female athletes at different competing levels—A pilot study. *BMC Sports Sci. Med. Rehabil.* **2019**, *11*, 1. [CrossRef] [PubMed]
10. Bonaiuto, V.; Gatta, G.; Rmagnoli, C.; Boatto, P.; Lanotte, N.; Annino, G. A Pilot Study on the e-Kayak System: A Wireless DAQ Suited for Performance Analysis in Flatwater Sprint Kayaks. *Sensors* **2020**, *20*, 542. [CrossRef] [PubMed]
11. Dansprint. Technical Information. Kayak Information. Available online: https://dansprint.com/dansprint-technical-specifikation/ (accessed on 6 December 2021).
12. van Someren, K.A.; Palmer, G.S. Prediction of 200-m sprint kayaking performance. *Can. J. Appl. Physiol.* **2003**, *28*, 505–517. [CrossRef] [PubMed]
13. van Someren, K.A.; Phillips, G.R.W.; Palmer, G.S. Comparison of physiological responses to open water kayaking and kayak ergometry. *Int. J. Sports Med.* **2000**, *21*, 200–204. [CrossRef] [PubMed]
14. Fleming, N.; Donne, B.; Fletcher, D.; Mahony, N. A biomechanical assessment of ergometer task specificity in elite flatwater kayakers. *J. Sports Sci. Med.* **2012**, *11*, 16–25. [PubMed]
15. Bishop, D.; Bonetti, D.; Spencer, M.; Bishop, D.; Bonetti, D.; Spencer, M. The effect of an intermittent, high-intensity warm-up on supramaximal kayak ergometer performance. *J. Sports Sci.* **2014**, *21*, 13–20. [CrossRef] [PubMed]
16. Kleshnev, V. Interpretation of the Force Curve. Rowing Biomechanics Newsletter. 2018. Available online: http://biorow.com/index.php?route=information/news/news&news_id=29 (accessed on 2 October 2019).
17. Petrone, N.; Isotti, A.; Guerrini, G. *Biomechanical Analysis Of Olympic Kayak Athletes during Indoor Paddling*; Springer: New York, NY, USA, 2006; pp. 413–418.
18. Michael, J.S.; Rooney, K.B.; Smith, R.M. The dynamics of elite paddling on a kayak simulator. *J. Sports Sci.* **2012**, *30*, 661–668. [CrossRef] [PubMed]
19. Begon, M.; Colloud, F.; Lacouture, P. Measurement of contact forces on a kayak ergometer with a sliding footrest-seat complex. *Sports Eng.* **2009**, *11*, 67–73. [CrossRef]

sensors

MDPI

Article

The Effect of Paddle Stroke Variables Measured by Trainesense SmartPaddle® on the Velocity of the Kayak

Antti Löppönen [1,2], Tomi Vänttinen [3], Marko Haverinen [1,4] and Vesa Linnamo [1,*]

1 Faculty of Sport and Health Sciences, University of Jyväskylä, 40014 Jyväskylä, Finland; antti.ej.lopponen@jyu.fi (A.L.); marko.haverinen@varala.fi (M.H.)
2 Physical Activity, Sports and Health Research Group, Department of Movement Sciences, KU Leuven, 3000 Leuven, Belgium
3 KIHU-Research Institute for Olympic Sports, 40700 Jyväskylä, Finland; tomi.vanttinen@kihu.fi
4 Varala Sports Institute, 33240 Tampere, Finland
* Correspondence: vesa.linnamo@jyu.fi; Tel.: +358-405044800

Abstract: (1) Background: This study aimed to compare key variables of paddle stroke measured by a commercial Trainesense SmartPaddle® against the strain-gauge shaft and investigate how these variables are associated with the velocity of the boat among national-level canoe polo players. (2) Methods: This study involved 14 Finnish national-level canoe polo players. The measurement protocol consisted of three different paddling velocities, which were performed in indoor swimming pools. The velocity of the boat was calculated based on the performance time measured with the laser photocell gate. Canoe polo equipment was used in the study and a SmartPaddle sensor was attached to the paddle blade. A strain-gauge paddle shaft was used as a reference method to examine the validity of SmartPaddle. (3) Results: The stroke rate, force production time, mean and maximal force measured with the strain-gauge paddle shaft correlated strongly (r = 0.84–0.95, $p < 0.01$) with SmartPaddle. However, the SmartPaddle overestimated the maximum force compared to the strain-gauge shaft. Stroke rate (r = 0.86, $p < 0.01$), mean force (r = 0.79, $p < 0.01$), maximal force (r = 0.78, $p < 0.01$) and total absolute impulse (r = 0.70, $p < 0.01$) correlated positively and force production time negatively (r = −0.76, $p < 0.01$) with the velocity of the boat. (4) Conclusions: We conclude that the SmartPaddle provides promising information on stroke key variables when compared to the strain-gauge paddle shaft. The SmartPaddle is a new and interesting tool for biomechanical research and daily kayaking coaching in real open water conditions. However, more research and algorithm development are needed before the SmartPaddle can be used in everyday coaching sessions in kayaking.

Keywords: kayaking; biomechanics; canoe polo; inertial measurement unit

Citation: Löppönen, A.; Vänttinen, T.; Haverinen, M.; Linnamo, V. The Effect of Paddle Stroke Variables Measured by Trainesense SmartPaddle® on the Velocity of the Kayak. *Sensors* **2022**, *22*, 938. https://doi.org/10.3390/s22030938

Academic Editor: Giuseppe Vannozzi

Received: 10 November 2021
Accepted: 23 January 2022
Published: 26 January 2022

1. Introduction

To create a forward acceleration for the boat, the paddler must win the resistance forces of water and air [1–4]. Force, power, and stroke rate are the most important factors influencing the velocity of the boat [5–9]. In addition, it is important to produce a "bell-shaped" force profile, which leads to a bigger impulse [6,10,11]. The paddle stroke is most effective when the blade is perpendicular to the water surface [2,12] and the maximum force is reached just before this [13,14]. Studies focused on the paddle trajectory have shown that more experienced paddlers are capable of more symmetrical paddling performance [15–17] and have wider trajectories than inexperienced paddlers [16].

Traditionally, the direction of the paddle stroke is parallel to the kayak's centerline, but on a wing paddle, the trajectory is slightly different because the blade's wing shape produces buoyancy compared to a traditional paddle [11,12,15]. When the paddler is using the wing paddle, the trajectory should be on a slightly more curved line, moving beyond the centerline of the boat [15]. This has been found to prolong the optimal duration of the propulsion generation position [11]. The stroke trajectory is thus one of the key factors related to paddling performance and varies between different types of paddling such as canoe polo and sprint kayaking. Although the canoe polo is paddled with a traditional paddle (flat-blade) and should follow the same principles as other kayak sports using a traditional paddle, the actual research evidence on canoe polo is still limited.

Most of the biomechanical kayaking studies have been conducted through motion analysis and direct force measurements. Studies have been made by ergometers in laboratory conditions [8,18–20] and open water environments [7,10,12,21–23]. Studies in an open water environment often use a strain-gauge paddle shaft [6,21,23] to measure the bending of the shaft and thereby the variables of the stroke. This method has also been found to be a reliable analyze on-water rowing performance [24]. Paddle blade trajectory-related studies have previously been done with an ergometer in laboratory conditions [20,25] and open water using cinematographic cameras [12,26].

However, the limitation of these methods is that they cannot be used to study the direction of the paddle stroke or the trajectory of the paddle blade when paddling in open water conditions for long distances (competitive analyzes). Moreover, many methods such as 3D-motion analysis are not suitable for direct transfer to an open water environment especially for daily (technical) coaching sessions, where a stand-alone system is often required.

The new compact sensor technology could offer new opportunities for biomechanical research and daily coaching in water sports. The small-sized inertial measurement units allow these devices to be used in real conditions [27–29]. Inertial measurement units are capable of measuring, processing, and storing data in a compact size and are also very light and are thus possible to integrate into equipment [27]. Inertial measurement units have been used in water sports especially in swimming [30,31] but also in kayaking to measure kayak movements [32,33]. It is also noteworthy that the development of algorithms plays a key role as sensor technology is a growing method in the biomechanical measurement of sports. For example in swimming, stroke rate, swimming velocity, and stroke phases have been the most important variables in algorithm development [34]. However, to our knowledge, the inertial measurement units have not been used to measure similar kayaking stroke variables such as force, stroke, or force production time. The inertia measurement unit could provide a tool to estimate the direction of paddle stroke under real conditions because the data measured by its sensors can be used to determine trajectories.

In this study, we use the Trainesense SmartPaddle® which is an advanced commercial wearable underwater sensor. It has been widely used in swimming to examine technique (trajectory, velocity, and orientation of the hand) and the applied forces during a swimming stroke [35]. This study aimed to compare key variables of paddle stroke measured by a SmartPaddle against the strain-gauge paddle shaft. In addition, we investigate how these key variables are associated with the velocity of the boat among national-level canoe polo players.

2. Materials and Methods

2.1. Study Design and Participants

This cross-sectional study consisted of two datasets. The first dataset (n = 6, all were men) compared the results of the SmartPaddle and a strain-gauge paddle shaft in key paddle stroke variables. The second study (total n = 14, men n = 12, women n = 2) investigated the effect of key paddle stroke variables measured on a SmartPaddle on the velocity of a boat.

The participants of this study were Finnish national-level canoe polo players (Table 1). The average age of the dataset one participants (n = 6) was 39.3 ± 10.1 years, height 183.3 ± 10.6 cm, and weight 81.3 ± 13.0 kg. The average angle of the blade was 80.0 ± 0.0° and the paddle length was 206.0 ± 0.0 cm. The average age of the dataset two participants (n = 14) was 40.8 ± 17.2 years, height 182.3 ± 7.6 cm, and weight 81.2 ± 11.6 kg. The average angle of the blade was 69.1 ± 17.9° and the paddle length was 206.4 ± 0.9 cm.

Table 1. The descriptive information of the study participants.

Mean (SD)	1st Dataset (n = 6) [1]	2nd Dataset (n = 14) [1]
Age [years]	39.3 ± 10.1	40.8 ± 17.2
Height [cm]	183.3 ± 10.6	182.3 ± 7.6
Weight [kg]	81.3 ± 12.9	81.2 ± 11.6
Blade angle [deg]	80.0 ± 0.0 [2]	69.1 ± 17.9
Paddle length [cm]	206.0 ± 0.0 [2]	206.4 ± 0.9

SD = Standard Deviation, [1] Descriptive table is missing data from five participants because they did not reply to the questionnaire sent afterward (1st dataset = 2 participants, 2nd dataset = 3 participants). [2] no adjustable.

The inclusion criterion for the participants were: (1) at least three years of experience of competition kayaking, (2) basic health, and (3) were not allowed to have any injuries or other problems that interfered with their kayaking performance. The study protocol followed the principles of the Declaration of Helsinki. The study was approved by the Human Sciences Ethics Committee of the University of Jyväskylä. All participants signed written informed consent before participating in the study.

2.2. Measurement

Measurements were performed in three different indoor swimming pools, in two locations in Finland. All participants in dataset one were measured in a 25-m pool with a measurement distance of 15 m. Four participants of dataset two were measured in a 50-m pool (measuring distance 30 m) and the remaining 10 participants were measured in a 25-m pool (measuring distance 15 m). Based on the paddling velocities, the measurement distance did not affect the results of the analysis. At a 30-m analysis distance (50-m pool), mean (SD) 13.9 ± 1.7 strokes per side were included in the analysis at velocity one, 14.6 ± 2.4 strokes at velocity two, and 15.5 ± 2.1 strokes at velocity three. At a 15-m analysis distance (25-m pool), 6.4 ± 0.8 strokes were included in the analysis at velocity one, 6.9 ± 1.0 strokes at velocity two, and 7.1 ± 0.9 strokes at velocity three. The velocity of the canoe was calculated based on the performance time measured with the laser photocell gates placed on the edges of the pool. The distance between the photocells was measured and the average velocity was calculated by dividing the paddled distance by the performance time ($v = s/t$, where v = velocity, s = distance, and t = time). The photocells were placed in the pools so that the boat could be accelerated (acceleration distance was 5 m) to the cruising velocity before the first photocell gate started the timing.

The measurement protocol for dataset one (SmartPaddle & strain-gauge shaft comparison) consisted of three different velocities of paddling performances. First participants were instructed to paddle at a basic endurance pace (velocity 1), which was described as a slow and calm but strong pace that can be maintained for a long time. The second velocity was clearly (velocity 2) faster paddling, which was described as strong, but technically pure, paddling. This velocity was performed with three different techniques (normal, energetic beating stroke, delayed long and light stroke) to add heterogeneity to the data used in the validity analysis. The last was maximum velocity (velocity 3), which was described as a near maximum performance, where the pace is almost at maximum, but the paddling technique remains high quality. In total, dataset one contained a total of 60 data points (=6 participants × 5 different performances/participant × 2 sides (left and right). The velocities one (1.95 ± 0.22 m/s) two (2.20 ± 0.21 m/s) and three (2.50 ± 0.20 m/s) differed significantly ($p < 0.05$) as planned. The velocities of the two different technical styles (energetic beating stroke 2.36 ± 0.26 m/s & delayed long and light stroke 2.25 ± 0.21 m/s) were the same as normal style of velocity two ($p = 0.313$, $p = 0.563$. respectively) as planned.

Dataset two consisted of the three different velocities with normal paddling performances mentioned above (velocity 1, 2 & 3), to study the effect of key paddle stroke variables measured on a SmartPaddle on the velocity of a boat. Dataset two contained a total of 84 data points (=14 participants × 3 different velocity performances/participant × 2 sides (left and right)). The velocity of the boat differed significantly between different velocities ($p < 0.001$) as planned.

2.3. Instrumentation

The commercial SmartPaddle (Trainesense Oy, Tampere, Finland) was used as a measuring unit (9-axis IMU + pressure sensor). The device uses a sampling frequency of 100 Hz and the data was stored in the Trainesense Analysis Center cloud service. The data was uploaded to the mobile phone and onward Analysis Center via a Bluetooth connection. The SmartPaddle was mounted on the back of the blade (Braca-Sport Polo Kinetic max, adjustable shaft, zero-length 205 cm) using silicone straps, and a 15 mm diameter hole was drilled in the paddle blade for the pressure sensor (Figure 1).

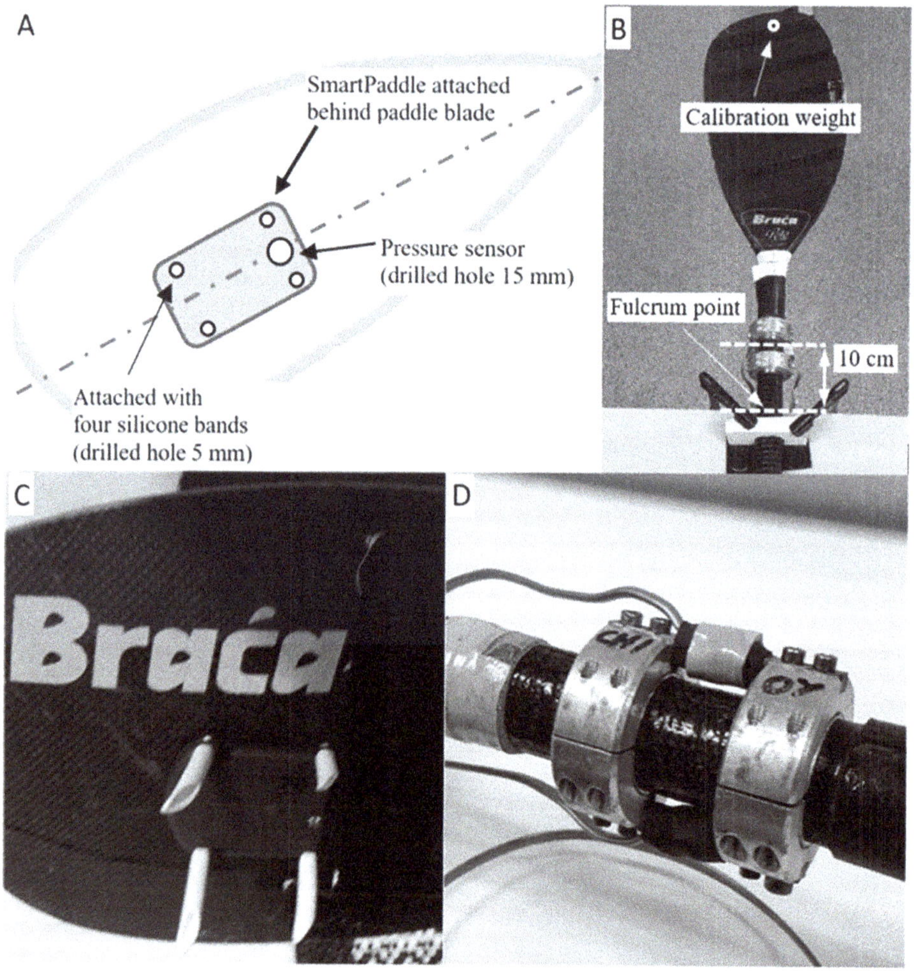

Figure 1. (**A**) SmartPaddle place on the blade, (**B**) Strain-gauge shaft in calibration board (without IMU attached), (**C**) SmartPaddle attached to the blade, (**D**) structure of the strain-gauge sensors.

A strain-gauge paddle shaft (self-manufactured, University of Jyväskylä, Finland) [9] was used for the comparison of the SmartPaddle (Figure 1D). This same strain-gauge equipment has also been used in previous research when examining the relationships between force-time parameters and kayak velocity [9]. The data produced by the strain-gauge sensors were amplified with separate amplifiers (TRtesti Oy, Jyväskylä Finland) and stored on a recordable AD converter (self-manufactured "LIIKE card", Sports Technology, University of Jyväskylä, Vuokatti, Finland) at a frequency of 250 Hz. The gauge paddle shaft was calibrated using 0 kg, 4 kg, and 10 kg weights to calculate the linear gain factor. During the calibration process, the strain gauge shaft was attached to the calibration table with the clamps and the fulcrum point was at the edge of the table (Figure 1B). The linear gain factor determined from the calibration was entered into the recordable AD Converter analysis software.

2.4. Data Processing

The SmartPaddle calculation algorithm has not been published. The SmartPaddle generates the processed data via the closed Matlab GUI (Graphical User Interface) (Tampere, Finland) developed by the Trainesense Oy (Tampere, Finland) meaning that there is no access to SmartPaddle raw data and the algorithm calculation constants can not be adjusted. This was also the reason why the comparisons between the strain-gauge shaft and the SmartPaddle used averaged variables as it was not possible to perform the actual stroke (SmartPaddle)-stroke (strain-gauge shaft) synchronization due to lack of the raw data from the SmartPaddle. In the present study, the SmartPaddle variables were normalized to paddle blade (Bracsa polo kinetic max, 735 cm^2) area and the data were processed separately for the right and left blades.

The mean force was the average total force for that time when the blade is in the water. The maximum force was the highest total (all directions) force that the SmartPaddle registered during the entire performance. The force production time was defined as the duration during which the blade produced force at more than 30% of the stroke maximum force. The absolute total impulse was total force multiplied by the whole time when the blade was under the water surface [35]. The SmartPaddle analysis consisted of only the cruising phase of the performances without the acceleration and braking phases. All variables except maximum force were averages of the strokes included in the analysis. The relative forward, lateral and vertical impulses estimated by the SmartPaddle were calculated in line with kayaking direction.

The data stored from the strain-gauge paddle shaft was processed using Matlab (R2020b, The MathWorks Inc., Natick, MA, USA). First, the raw data were filtered using a moving-average filter (movmean function) with a value of 10 data points (filter window) [36], to attenuate the largest interference peaks. Then the maximum force (max–function) of the paddle stroke was determined from the filtered data by identifying a peak force value (highest force) of the entire performance (cruising phase). Finally, mean stroke force and force production time were calculated using the same 30% force limit as the SmartPaddle algorithm in order to make the analysis comparable. This was done by determining a level of 30% of the maximum force. The duration and mean of the force production time were determined from the force vector above this level. The left and right sides were processed separately in the analysis, with both sides forming their own data points in the statistical analysis. Mean values (except maximum force) of total kayaking performances (cruising phase) were used in this strain-gauge paddle shaft analysis to make it comparable with the SmartPaddle analysis.

2.5. Statistical Analysis

Results are reported as mean and standard deviation (SD). Shapiro-Wilk normality test and histograms visual evaluation (normal distribution curve and distribution skewness and kurtosis) were used to check the normality of the data. The Shapiro-Wilk test indicated that some of the variables were not normally distributed (stroke rate, boat velocity, lateral and vertical relative impulse). Similarly, the visual interpretation of the histograms and the small sample size (n = 6 & n = 14) supported the selection of non-parametric tests. Associations between the data measured by the SmartPaddle and the strain-gauge paddle shaft were tested with the Spearman correlation coefficient. The correspondence between the data measured by the SmartPaddle and the gauge-paddle shaft was evaluated with two-way random (absolute agreement), single measure intra-class correlation coefficients (ICC) [37], and visualized with Bland-Altman plots [38]. ICC was used to characterize the correspondence as poor (<0.40), fair (0.40 to <0.60), good (0.60 to <0.75), or excellent ($\geq$0.75) [39]. The relationship between velocity and key paddle variables measured with the SmartPaddle was also studied by the Spearman correlation coefficients (two-tailed). Differences between velocities groups and strain-gauge paddle shaft and the SmartPaddle were tested using related–samples Wilcoxon signed-rank test. Statistical significance was set at $p \leq 0.05$ (two-tailed) and analyses were performed with the R statistical computing software (version 4.0.3) [40].

3. Results

3.1. Strain-Gauge Shaft Comparison

The SmartPaddle measured higher maximum force values than the strain-gauge paddle shaft ($p < 0.001$) (Table 2). The strain-gauge paddle shaft maximum force was 125.4 $\pm$ 34.2 N, while the mean of the maximum force of the SmartPaddle was 152.1 $\pm$ 57.4 N. However, the correlation between SmartPaddle and strain-gauge paddle shaft was strong 0.84 ($p < 0.001$), and the ICC was classified to be good 0.64 (Confidence Interval CI 95% 0.21, 0.82) ($p = 0.003$). The strain-gauge paddle shaft mean force was 85.4 $\pm$ 25.3 N and the SmartPaddle 83.3 $\pm$ 30.3 N, with a strong correlation of 0.88 ($p < 0.001$) and an excellent ICC of 0.86 (CI, 0.78, 0.92) ($p < 0.001$). This difference in mean force was not statistically significant between these two methods ($p = 0.083$).

Table 2. Comparison between the strain-gauge paddle shaft and the SmartPaddle (9-axis IMU+ pressure sensor) in key paddle stroke variables ($n = 6$).

Mean (SD)	Strain-Gauge	SmartPaddle	*p*-Value [1]	r [2]	ICC [3]
Maximal force [N]	125.4 ± 34.2	152.1 ± 57.4	<0.001	0.84 **	0.64 ***
Mean force [N]	85.4 ± 25.3	83.3 ± 30.3	=0.083	0.88 **	0.86 ***
Force prod. time [s]	0.43 ± 0.08	0.44 ± 0.10	=0.472	0.88 **	0.81 ***
Stroke rate [1/min]	84.8 ± 25.6	80.7 ± 21.0	=0.199	0.95 **	0.95 ***

SD = Standard deviation, [1] Wilcoxon signed rank test, [2] Spearman's rank correlation coefficient, [3] ICC = Intraclass Correlation Coefficient, ** Correlation is statistically significant 0.01 level (2-tailed), *** $p < 0.001$.

The force production time strongly correlated between the strain-gauge paddle shaft and the SmartPaddle (0.88, $p < 0.01$) and the average difference was only 0.01 s ($p = 0.472$) The ICC was classified to be excellent 0.81 (CI, 0.70, 0.88) ($p < 0.001$). Of all the variables, the strongest correlation was found in the stroke rate 0.95 ($p < 0.01$) and it did not differ statistically ($p = 0.199$), and the ICC was excellent at 0.95 (CI, 0.89, 0.98) ($p < 0.001$) (Table 2). Scatter plots and R^2 values between SmartPaddle and strain-gauge shaft in key paddle stroke variables are presented in Figure 2.

The Bland-Altman analysis (Figure 3) showed that there were only a few cases outside the concordance limits (95%) and no systematic error difference was noticeable, as the measurement results were evenly distributed on both sides of the zero level.

3.2. Key Variables to Velocity

Stroke rate ($r = 0.86$, $p < 0.001$) and mean force ($r = 0.79$, $p < 0.001$) were most strongly related to boat velocity and differed between all kayaking velocities ($p < 0.001$) (Tables 3 and 4). The absolute impulse ($r = 0.70$, $p < 0.001$) was also positively related to the velocity of the boat and differed significantly ($p < 0.001$) between the velocities while the force production-time was negatively related ($r = -0.76$, $p < 0.001$) to the velocity of the boat.

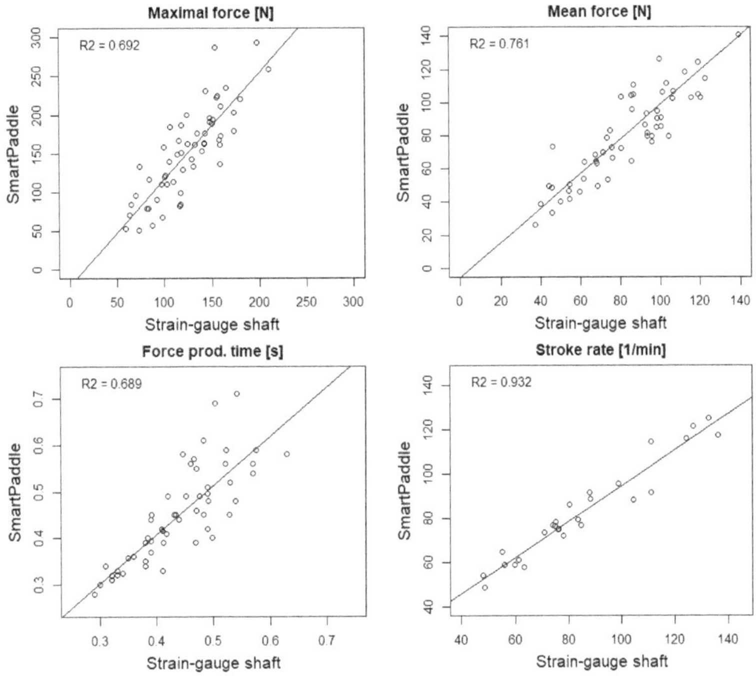

Figure 2. Scatter plots and R^2 values between SmartPaddle and strain-gauge shaft in key paddle stroke variables (n = 6) (R^2 = coefficient of determination showing how close the data are to the fitted regression line).

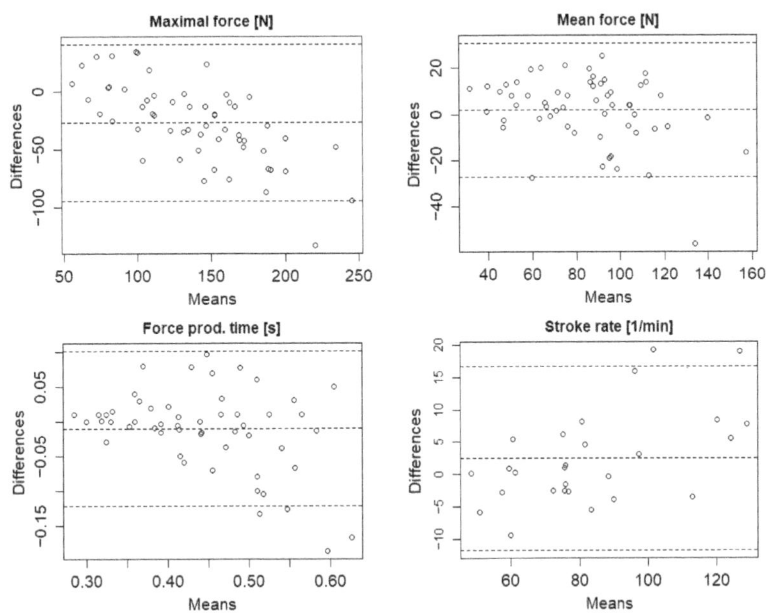

Figure 3. Bland–Altman analysis between SmartPaddle and strain-gauge shaft in key paddle stroke variables (n = 6).

Table 3. Descriptive characteristics of the participants and results of the testing velocities (n = 14).

Mean (SD)	Velocity 1	Velocity 2	Dif V1–V2 [1]	Velocity 3	Dif V2–V3 [1]
Velocity [m/s]	2.1 ± 0.2	2.4 ± 0.1	<0.001	2.7 ± 0.3	<0.001
Maximal force [N]	112.6 ± 38.4	145.1 ± 40.1	<0.001	224.9 ± 67.8	<0.001
Mean force [N]	61.6 ± 17.9	83.6 ± 21.8	<0.001	115.8 ± 32.2	<0.001
Force prod. time [s]	0.46 ± 0.06	0.40 ± 0.06	<0.001	0.34 ± 0.07	<0.001
Stroke rate [1/min]	75.7 ± 14.3	93.5 ± 13.4	<0.001	119.3 ± 19.7	<0.001
Total impulse abs [Ns]	58.6 ± 16.8	69.2 ± 15.8	<0.001	90.1 ± 22.8	<0.001
Impulse forward [%] [2]	49.50 ± 4.29	50.36 ± 4.23	0.387	49.75 ± 4.05	0.770
Impulse lateral [%] [2]	28.81 ± 4.90	29.60 ± 5.10	0.058	32.08 ± 5.14	0.036
Impulse vertical [%] [2]	21.69 ± 4.45	20.04 ± 3.81	0.002	18.17 ± 4.24	0.009

SD = Standard deviation, [1] Related-samples Wilcoxon signed-rank test, [2] Portion of total impulse.

Table 4. The Spearman's rank correlation coefficient by velocity groups (n = 14).

	Velocity 1	Velocity 2	Velocity 3	All Velocity
Maximal force [N]	0.60 **	0.56 **	0.74 **	0.78 **
Mean force [N]	0.61 **	0.47 *	0.73 **	0.79 **
Force prod. time [s]	−0.54 **	−0.40 *	−0.56 **	−0.76 **
Stroke rate [1/min]	0.82 **	0.34	0.76 **	0.86 **
Total impulse abs [Ns]	0.43 *	0.52 **	0.74 **	0.70 **
Impulse forward [%]	−0.15	−0.01	−0.18	−0.09
Impulse lateral [%]	0.45 *	0.11	0.58 **	0.42 **
Impulse vertical [%]	−0.27	−0.07	−0.46 *	−0.42 **

* $p < 0.05$, ** $p < 0.01$ (two-tailed).

The relative lateral impulse was positively and moderately related to boat velocity (r = 0.42, $p < 0.001$) while the vertical impulse was negatively and moderately related to velocity (r = −0.42, $p < 0.001$). There was no change in the forward impulse as the velocity increased and it did not correlate significantly with the velocity (r = −0.09, $p = 0.429$).

4. Discussion

The present study suggests that the Trainesense SmartPaddle® offers promising opportunities to measure paddle stroke key variables acknowledging that the calculation algorithms and device mounting need to be further developed more suitable for kayaking in the future. The stroke rate, force, force-time, and total absolute impulse were found to be key variables affecting the velocity of the boat among canoe polo players and these results were also well in line with previous studies with sprint kayakers [5,8]. Moreover, the Trainesense SmartPaddle® made it possible to study the direction of the stroke, and in this study, when the velocity of the boat increased, the relative amount of the lateral impulse of the paddle stroke increased, while the vertical decreased. The Trainesense SmartPaddle can provide new insights for technical coaching in kayaking in real open water conditions.

In this study, good to excellent agreement was found between the SmartPaddle and the strain-gauge shaft in key variables of paddle stroke (0.64–0.95, $p < 0.001$). The absolute differences between the SmartPaddle and the reference method in force production time, mean force, and stroke rate were small. The SmartPaddle overestimated maximum force compared to a strain-gauge shaft ($p > 0.001$) and the agreement was also lowest in this variable (0.64, $p < 0.001$). This may be due to differences in data processing, especially data filtering. We used a moving average (movmean) filter to analyze the strain-gauge shaft data, which may cut the highest force peaks more than the filtering algorithm used by the SmartPaddle. However, it was more likely that the SmartPaddle overestimated the highest values such as maximum force and absolute impulse due to issues related to their hardware and algorithms. This was indicated by the sensitivity analysis between the left and right sides which suggested that it was indeed the SmartPaddle that overestimated the maximum force compared to the strain-gauge shaft (Supplementary Table S1). Unfortunately, this difference could not be studied further because the algorithm used by the SmartPaddle is a trade secret and therefore these estimates should be treated with caution.

To our knowledge, inertial measurement units have not previously been used to measure the trajectory of the paddle in a real open water environment. Furthermore, to our knowledge, this is the first study to measure kayak paddling technique under real conditions through a paddle attached IMU sensor. The advantage of SmartPaddle is that it can be used to identify the direction of the paddle stroke. The results of this study suggested that as the velocity of the boat increased, the relative amount of lateral impulse of the paddle stroke increased, vertical decreased, and no difference was observed in the relative amount of forward impulse. This observation suggests that when boat velocity increases, paddles tend to produce more force by changing the trajectory of the paddle blade. On the other hand, this may be affected by the fact that the stroke is shorter at a higher stroke rate [6,10] and thus the vertical impulse is lower. This observation has been found in the previous studies which have shown that as the stroke rate becomes sufficiently high, the duration of the water phase also shortens, especially from the beginning phase of the stroke (first 1/3) and the end phase (last 1/3) of the stroke [21]. However, it should be noted that this study used a standard blade (canoe polo blade) which has been found to differ from the wing paddle trajectory [15]. The results of the left and right sides comparison indicated (Supplementary Table S2) that there was a slight difference in the direction of the strokes between the sides which is in agreement with the previous studies in which asymmetries between left and right strokes trajectories [6,20,33,41].

Although the results provide indications that the SmartPaddle, which is widely used in swimming, can be used to measure the biomechanics of kayaking, more studies on sensor reliability and validity are still needed. Furthermore, the SmartPaddle appears to work well on key variables of paddle stroke based on these results, but the trajectory or direction of the paddle blade has not been validated in this or previous studies. This needs to be done in the future, especially if SmartPaddle is used in research. It is especially important to study how the sensor can measure variables on the wing paddle as one possible limitation may be the turbulent vortex [11,42] in the blade surface. In addition, further research is needed to find out why the SmartPaddle seemed to overestimate the maximum force compared to the strain-gauge shaft.

It has previously been found that there is a strong correlation between the average velocity of the kayak and the total impulse, regardless of the frequency, which would suggest that only by increasing the total impulse can the velocity of the kayak be increased [10]. In the future, it would be valuable to study the relationship between the forward impulse and the stroke rate to the kayak velocity. This could serve as one of the important efficiency indicators in kayaking, because this analysis makes it possible to estimate the proportion of the forward impulse.

Some limitations need to be kept in mind when interpreting the findings. Generalization of the results is restricted, particularly for sprint and marathon kayaking, as they exclusively use wing blades and the measurements in this study were performed on canoe polo equipment. In addition, the subjects' average age in this study was high (40.8 years) and standard deviation large (±17.2 years), which also impairs generalization of the results for younger international level athletes. It should be noted that the distances in this study were shorter than in the actual exercises, and for successful use in them SmartPaddle attachment to the paddle blade must be developed in the future and possibly integrated into the structure of the blade.

The strength of this study was that the measurement protocol was successful, the velocities were significantly different as was planned, and the paddling technique was pure. Conditions were standardized in the swimming pools so that pool depth, water temperature, or weather conditions did not affect the results. This study focused on the cruising velocities, as it is known from previous research [43] that the first accelerator paddle strokes differ from the cruising velocity paddle strokes and thus the data processing was more reliable. Participants in the study were experienced kayakers and their paddling techniques were well established. Thus, the performances were clean, and their technique was stable throughout the performance, although it could not be fully standardized due to individual differences. In this study, subjects did not report problems even though the device was attached to the paddle blade.

The SmartPaddle could support traditional video coaching in kayaking and strengthening the vision of coaches and athletes. If the development of the algorithm is continued, it should be done according to the requirements of kayaking, so that the device can operate reliably in training sessions and competition analyses to evaluate kayak paddling techniques in all types of kayak sports.

5. Conclusions

The SmartPaddle, consisting of a 9-axis IMU and pressure sensor, offers promising opportunities to measure stroke key variables when compared to the strain-gauge paddle shaft. Stroke rate, force, force production time, and total impulse were the stroke key variables, which affect the velocity of the kayak. The results of the study give preliminary indications that when the velocity of the kayak increases, the relative amount of lateral impulse increases, and the vertical impulse decreases while the forward impulse remains the same. Although kayaking has been extensively studied using motion analysis and force sensors, the study of the direction of paddle stroke has been done less. The results presented in this study suggest that the SmartPaddle can be attached to the paddle blade and thus provide a new way to study the direction of the paddle stroke in real open water conditions. However, more research and algorithm development are needed, especially to determine the reliability of the trajectory of the blade and the direction of stroke.

Supplementary Materials: The following are available online at https://www.mdpi.com/article/10.3390/s22030938/s1, Table S1: Left- and right-side comparison between strain-gauge shaft and SmartPaddle; Table S2: Left- and right-side comparison between velocities.

Author Contributions: Conceptualization, A.L., M.H., T.V. and V.L.; methodology, A.L., M.H., T.V. and V.L.; formal analysis, A.L.; writing—original draft preparation, A.L.; supervision, T.V. and V.L.; writing—review and editing, A.L., M.H., T.V. and V.L.; All authors have read and agreed to the published version of the manuscript.

Funding: This research was funded by the University of Jyväskylä, KIHU-Research Institute for Olympic Sports, Varala Sports Institute, and Finnish Canoeing and Rowing Federation.

Institutional Review Board Statement: The study was conducted according to the guidelines of the Declaration of Helsinki and approved by the Ethics committee of the University of Jyväskylä on 17 June 2019.

Informed Consent Statement: Informed consent was obtained from all subjects involved in the study.

Data Availability Statement: Pseudonymized datasets are available to external collaborators subject to agreement on the terms of data use and publication of results. To request the data, please contact Antti Löppönen (antti.ej.lopponen@jyu.fi).

Acknowledgments: The authors thank Trainesense Oy & Ari Auvinen for technical support during data processing. In addition, the authors thank laboratory engineer Sami Vierola from the Research Institute for Olympic Sports, who helped with the implementation of the measuring equipment.

Conflicts of Interest: The authors declare no conflict of interest.

References

1. Baker, J. The Evaluation of Biomechanic Performance Related Factors and On-Water Tests. *Int. Semin. Kayak-Canoe Coach. Sci.* **1998**, 50–66.
2. Michael, J.S.; Smith, R.; Rooney, K.B. Determinants of kayak paddling performance. *Sports Biomech.* **2009**, *8*, 167–179. [CrossRef] [PubMed]
3. Robinson, M.G.; Holt, L.E.; Pelham, T.W. The technology of sprint racing canoe and kayak hull and paddle designs. *Int. Sports J.* **2002**, *6*, 68–85.
4. Seiler, S. A biomechanical review of factors affecting rowing performance: Commentary. *Br. J. Sports Med.* **2002**, *36*, 402.

5. Brown, M.B.; Lauder, M.; Dyson, R. Notational analysis of sprint kayaking: Differentiating between ability levels. *Int. J. Perform. Anal. Sport* **2011**, *11*, 171–183. [CrossRef]
6. MacDermid, P.W.; Osborne, A.; Stannard, S.R. Mechanical work and physiological responses to simulated flat water Slalom Kayaking. *Front. Physiol.* **2019**, *10*, 260. [CrossRef]
7. McDonnell, L.K.; Hume, P.A.; Nolte, V. Place time consistency and stroke rates required for success in K1 200-m sprint kayaking elite competition. *Int. J. Perform. Anal. Sport* **2013**, *13*, 38–50. [CrossRef]
8. Tornberg, B.; Håkansson, P.; Svensson, I.; Wollmer, P. Forces applied at the footrest during ergometer kayaking among female athletes at different competing levels—A pilot study. *BMC Sports Sci. Med. Rehabil.* **2019**, *11*, 1. [CrossRef]
9. Mononen, H.; Viitasalo, J. Stroke parameters and kayak speed during 200 m kayaking. *Congr. Int. Soc. Biomech.* **1995**, 632–633.
10. Gomes, B.; Ramos, N.; Conceição, F.; Sanders, R.; Vaz, M.; Vilas-Boas, J.P. Paddling force profiles at different stroke rates in elite sprint kayaking. *J. Appl. Biomech.* **2015**, *31*, 258–263. [CrossRef]
11. Sanders, R.H.; Baker, J.D. Evolution of technique in flatwater kayaking. *Sci. Pract. Canoe* **1998**, 67–81.
12. Mann, R.V.; Kearney, J.T. A biomechanical analysis of the olympic-style flatwater kayak stroke. *Med. Sci. Sports Exerc.* **1980**, *12*, 183–188. [CrossRef] [PubMed]
13. Aitken, D.A.; Neal, R.J. An On-Water Analysis System for Quantifying Stroke Force Characteristics during Kayak Events. *Int. J. Sport Biomech.* **2016**, *8*, 165–173. [CrossRef]
14. Fleming, N.; Donne, B.; Fletcher, D.; Mahony, N. A biomechanical assessment of ergometer task specificity in elite flatwater kayakers. *J. Sports Sci. Med.* **2012**, *11*, 16–25.
15. Kendal, S.J.; Sanders, R.H. The Technique of Elite Flatwater Kayak Paddlers Using the Wing Paddle. *Int. J. Sport Biomech.* **1992**, *8*, 233–250. [CrossRef]
16. Limonta, E.; Squadrone, R.; Rodano, R.; Marzegan, A.; Veicsteinas, A.; Merati, G.; Sacchi, M. Tridimensional kinematic analysis on a kayaking simulator: Key factors to successful performance. *Sport Sci. Health* **2010**, *6*, 27–34. [CrossRef]
17. Helmer, R.J.N.; Farouil, A.; Baker, J.; Blanchonette, I. Instrumentation of a kayak paddle to investigate blade/water interactions. *Procedia Eng.* **2011**, *13*, 501–506. [CrossRef]
18. Bjerkefors, A.; Tarassova, O.; Rosén, J.S.; Zakaria, P.; Arndt, A. Three-dimensional kinematic analysis and power output of elite flat-water kayakers. *Sports Biomech.* **2018**, *17*, 414–427. [CrossRef]
19. Lee, C.H.; Nam, K.J. Analysis of the kayak forward stroke according to skill level and knee flexion angle. *Int. J. Bio-Sci. Bio-Technol.* **2012**, *4*, 41–48.
20. Therrien, M.; Colloud, F.; Begon, M. Effect of stroke rate on paddle tip path in kayaking. *Mov. Sport. Sci.-Sci. Mot.* **2012**, *75*, 113–120. [CrossRef]
21. Gomes, B.B.; Ramos, N.V.; Conceição, F.; Sanders, R.; Vaz, M.; Vilas-Boas, J.P. Paddling time parameters and paddling efficiency with the increase in stroke rate in kayaking. *Sports Biomech.* **2020**, 1–9. [CrossRef] [PubMed]
22. Kong, P.W.; Tay, C.S.; Pan, J.W. Application of instrumented paddles in measuring on-water kinetics of front and back paddlers in k2 sprint kayaking crews of various ability levels. *Sensors* **2020**, *20*, 6317. [CrossRef] [PubMed]
23. Macdermid, P.W.; Gilbert, C.; Jayes, J. Using a kayak paddle power-meter in the sport of whitewater slalom. *J. Hum. Sport Exerc.* **2019**, *15*, 105–118. [CrossRef]
24. Lintmeijer, L.L.; Onneweer, J.P.T.; Hofmijster, M.J.; Wijgergangs, W.A.; De Koning, H.; Clairbois, B.; Westerweel, J.; Grift, E.J.; Tummers, M.J.; Van Soest, A.J. Towards determination of power loss at a rowing blade: Validation of a new method to estimate blade force characteristics. *PLoS ONE* **2019**, *14*, e0215674. [CrossRef] [PubMed]
25. Bjerkefors, A.; Rosén, J.S.; Tarassova, O.; Arndt, A. Three-Dimensional Kinematics and Power Output in Elite Para-Kayakers and Elite Able-Bodied Flat-Water Kayakers. *J. Appl. Biomech.* **2019**, *35*, 93–100. [CrossRef] [PubMed]
26. Plagenhoef, S. Biomechanical analysis of olympic flatwater kayaking and canoeing. *Res. Q. Am. Alliance Health Phys. Educ. Recreat. Danc.* **1979**, *50*, 443–459. [CrossRef]
27. Espinosa, H.G.; Lee, J.; James, D.A. The Inertial Sensor: A Base Platform for Wider Adoption in Sports Science Applications. *J. Fit. Res.* **2015**, *4*, 13–20.
28. Magalhaes, F.A.d.; Vannozzi, G.; Gatta, G.; Fantozzi, S. Wearable inertial sensors in swimming motion analysis: A systematic review. *J. Sports Sci.* **2015**, *33*, 732–745. [CrossRef]
29. Victorino, M.N.; Jiang, X.; Menon, C. Wearable Technologies and Force Myography for Healthcare. In *Wearable Technology in Medicine and Health Care*; Elsevier (Academic Press): Amsterdam, The Netherlands, 2018; pp. 135–152.
30. Ramos Félix, E.; da Silva, H.P.; Olstad, B.H.; Cabri, J.; Lobato Correia, P. SwimBIT: A Novel Approach to Stroke Analysis during Swim Training Based on Attitude and Heading Reference System (AHRS). *Sports* **2019**, *7*, 238. [CrossRef]
31. Li, R.; Cai, Z.; Lee, W.; Lai, D.T. A wearable biofeedback control system based body area network for freestyle swimming. In Proceedings of the 2016 38th Annual International Conference of the IEEE Engineering in Medicine and Biology Society (EMBC), Orlando, FL, USA, 16–20 August 2016; Volume 2016, pp. 1866–1869. [CrossRef]
32. Bifaretti, S.; Bonaiuto, V.; Federici, L.; Gabrieli, M.; Lanotte, N. E-kayak: A Wireless DAQ System for Real Time Performance Analysis. *Procedia Eng.* **2016**, *147*, 776–780. [CrossRef]
33. Bonaiuto, V.; Gatta, G.; Romagnoli, C.; Boatto, P.; Lanotte, N.; Annino, G. A pilot study on the e-kayak system: A wireless DAQ suited for performance analysis in flatwater sprint kayaks. *Sensors* **2020**, *20*, 542. [CrossRef] [PubMed]

34. Mooney, R.; Corley, G.; Godfrey, A.; Quinlan, L.R.; ÓLaighin, G. Inertial sensor technology for elite swimming performance analysis: A systematic review. *Sensors* **2015**, *16*, 18. [CrossRef] [PubMed]
35. Trainesense, O. SmartPaddle Data Explained Trainesense. Available online: http://34.251.210.97/wp-content/uploads/2019/05/SmartPaddle_data_explained_short.pdf (accessed on 10 January 2021).
36. Smith, S.W. Moving Average Filters. In *Teoksessa Digital Signal Processing*; Elsevier (Newnes): Amsterdam, The Netherlands, 2003; pp. 277–284, ISBN 978-0-7506-7444-7.
37. Mehta, S.; Bastero-Caballero, R.F.; Sun, Y.; Zhu, R.; Murphy, D.K.; Hardas, B.; Koch, G. Performance of intraclass correlation coefficient (ICC) as a reliability index under various distributions in scale reliability studies. *Stat. Med.* **2018**, *37*, 2734–2752. [CrossRef] [PubMed]
38. Martin Bland, J.; Altman, D.G. Statistical methods for assessing agreement between two methods of clinical measurement. *Lancet* **1986**, *327*, 307–310. [CrossRef]
39. Cicchetti, D.V. Guidelines, Criteria, and Rules of Thumb for Evaluating Normed and Standardized Assessment Instruments in Psychology. *Psychol. Assess.* **1994**, *6*, 284–290. [CrossRef]
40. R Core Team. *R: A Language and Environment for Statistical Computing*; R Foundation for Statistical Computing: Vienna, Austria, 2020. Available online: https://www.R-project.org/ (accessed on 15 December 2021).
41. Baker, J.; Rath, D.; Sanders, R.; Kelly, B. A three dimensional analysis of male and female elite sprint kayak paddlers. In Proceedings of the XVIIth International Symposium on Biomechanics in Sports, Perth, Australia, 30 June–6 July 1999; pp. 53–56.
42. Jackson, P.S. Performance prediction for olympic kayaks. *J. Sports Sci.* **1995**, *13*, 239–245. [CrossRef]
43. Gomes, B.; Viriato, N.; Sanders, R.; Conceição, F.; Vaz, M. Analysis of the on-water paddling force profile on an elite kayaker. *Port. J. Sport Sci.* **2011**, *11*, 259–262.

sensors

MDPI

Article

Framework for In-Field Analyses of Performance and Sub-Technique Selection in Standing Para Cross-Country Skiers

Camilla H. Carlsen [1,*], Julia Kathrin Baumgart [1], Jan Kocbach [1,2], Pål Haugnes [1], Evy M. B. Paulussen [1,3] and Øyvind Sandbakk [1]

[1] Centre for Elite Sports Research, Department of Neuromedicine and Movement Science, Faculty of Medicine and Health Sciences, Norwegian University of Science and Technology, 7491 Trondheim, Norway; julia.k.baumgart@ntnu.no (J.K.B.); jan.kocbach@ntnu.no (J.K.); pal.haugnes@ntnu.no (P.H.); Evy.paulussen@mumc.nl (E.M.B.P.); oyvind.sandbakk@ntnu.no (Ø.S.)
[2] NORCE Norwegian Research Centre AS, 5008 Bergen, Norway
[3] Faculty of Health, Medicine & Life Sciences, Maastricht University, 6200 MD Maastricht, The Netherlands
* Correspondence: camilla.h.carlsen@ntnu.no; Tel.: +47-452-40-788

Abstract: Our aims were to evaluate the feasibility of a framework based on micro-sensor technology for in-field analyses of performance and sub-technique selection in Para cross-country (XC) skiing by using it to compare these parameters between elite standing Para (two men; one woman) and able-bodied (AB) (three men; four women) XC skiers during a classical skiing race. The data from a global navigation satellite system and inertial measurement unit were integrated to compare time loss and selected sub-techniques as a function of speed. Compared to male/female AB skiers, male/female Para skiers displayed 19/14% slower average speed with the largest time loss (65 ± 36/35 ± 6 s/lap) found in uphill terrain. Female Para/AB skiers utilized DP, DK, and DIA, 61/43%, 15/10%, and 25/47% of the distance at low speeds, respectively, while the corresponding numbers for male Para/AB skiers were 58/18%, 1/13%, and 40/69%. At higher speeds, female Para/AB skiers utilized DP and OTHER, 26/52% and 74/48% of the distance, respectively, while corresponding numbers for male Para/AB skiers were 29/66% and 71/34%. This indicates different speed thresholds of the classical sub-techniques for Para than AB skiers. The framework provides a point of departure for large-scale international investigations of performance and related factors in Para XC skiing.

Keywords: micro-sensor technology; GNSS; IMU; disability; heterogenous group; cross-country skiing race; performance analysis; sub-technique classification; time factor

Citation: Carlsen, C.H.; Kathrin Baumgart, J.; Kocbach, J.; Haugnes, P.; Paulussen, E.M.B.; Sandbakk, Ø. Framework for In-Field Analyses of Performance and Sub-Technique Selection in Standing Para Cross-Country Skiers. *Sensors* **2021**, *21*, 4876. https://doi.org/10.3390/s21144876

Academic Editor: Vesa Linnamo

Received: 17 June 2021
Accepted: 14 July 2021
Published: 17 July 2021

Publisher's Note: MDPI stays neutral with regard to jurisdictional claims in published maps and institutional affiliations.

1. Introduction

Para cross-country (XC) skiing is a winter sport performed by skiers with different disabilities. Depending on their disability, Para XC skiers compete in three categories, which are further divided into classes, based on the functional impact of the disability on XC skiing performance: (1) physically impaired sitting skiers (classes: LW10–12), (2) physically impaired standing skiers (classes: LW2–9), and (3) visually impaired standing skiers (classes: B1–3) [1,2]. Additionally, within each category, a class-specific time factor is used to calculate the final race time [1,2].

Physically impaired standing XC skiers constitute a heterogenous group of skiers with different disabilities, which range from having an amputation to muscle weakness or loss of muscle control [1,2]. Similar to able-bodied (AB) XC skiers, standing Para XC skiers compete within the classical and skating styles in race courses consisting of undulating terrain with uphill, flat, and downhill segments [3]. The varying terrain during XC skiing races leads to substantial variation in speed, which is regulated by selection of pacing strategies, sub-techniques, and related kinematic patterns [4–7]. In the classical style, XC skiers alternate between double poling (DP), which is used at higher speeds on a wide range of inclines [8,9], kick double poling (DK), which is used at moderate speeds in the transition between different terrains [10], diagonal stride (DIA), which is primarily used at low speeds in moderate to steep uphill terrain [11,12], and the herringbone technique (HRB), which is used at low speeds in very steep uphill terrain [13]. During downhill sections, the skiers employ the tuck position without pole and leg actions, and various turn techniques are adapted to manage turning [4,14]. The choice of sub-technique and regulation of kinematic patterns is complex and influenced by individual preferences, internal, and external factors [6,11,12,15–18]. In classical AB XC skiing, it has been suggested that there are speed [6,17,18] and incline [11,12,16] thresholds for the use of the sub-techniques. Additionally, the skiers' physical capacity will influence the speed and choice of sub-technique [19,20]. In this context, the ability to use the different sub-techniques may additionally be dependent on functional limitations related to the individual disability among standing Para XC skiers. Accordingly, the sub-technique selected at different speeds may differ between standing Para and AB XC skiers.

Related to determination of the above parameters, micro-sensor technology has allowed detailed in-field performance analyses with continuous speed and time tracking, as well as automatic sub-technique classification, and is widely used among AB XC skiers [6,15,21]. However, in standing Para XC skiing, analyses of in-field performance or sub-technique distribution have not yet been done. Accordingly, a framework for such analyses would be beneficial for providing new insights into the technical and tactical aspects, as well as the effect of terrain and external conditions on the time factor, related to standing Para XC skiing performance.

Therefore, the aim of this study was to evaluate the feasibility of a framework based on micro-sensor technology for detailed analyses of in-field performance and sub-technique selection in Para XC skiing by using this framework in case-series to descriptively compare performance-related parameters between elite standing Para and AB XC skiers during a classical skiing race.

2. Materials and Methods

2.1. Participants

Three elite standing Para XC skiers (two male B3 skiers, one female LW4 skier) of the Norwegian national team, and nine elite AB XC skiers (five men, four women) of the Norwegian B national team participated in the study (Table 1). The male B3 skiers had 10% vision and were accompanied by a personal guide during the race. The female LW4 skier had linear scleroderma with reduced leg length, joint mobility, muscle mass, and strength in the one leg. Due to a limited number of elite standing Para XC skiers within the same category, AB XC skiers were used as reference to evaluate the feasibility of the framework. Among the male AB XC skiers, there was one participant with missing data due to complications with tracking during the race and one with an unfinished race. Their data were omitted and data of three male and four female AB XC skiers were included in the analyses. All participants signed an informed consent form and were made aware that they could withdraw from the study at any point without providing an explanation. The study was approved by the Norwegian Centre for Research Data (ID 49865/3/IJJ) and conducted in line with the Declaration of Helsinki.

Table 1. Age, body-mass, and training volume (mean ± SD) of the three Para and seven able-bodied (AB) XC skiers included in the analyses.

Parameter	Paralympic		Able-Bodied	
	Men (n = 2)	Woman (n = 1)	Men (n = 3)	Women (n = 4)
Age (years)	24.0 ± 2.8	19.0	25.0 ± 1.5	23.5 ± 1.3
Body-mass (kg)	70.5 ± 1.9	61.0	83.0 ± 2.0	63.5 ± 4.1
Training volume (hours·week^{-1})	13.5 ± 5.0	11.0	16.2 ± 1.0	14.5 ± 1.0

2.2. Design

During a national competition, participants performed a time-trial XC skiing race on snow using the classical style. The time-trial was performed on a 2.5 km race course, where female Para and AB XC skiers raced 10 km (4 × 2.5 km) and male Para and AB XC skiers raced 15 km (6 × 2.5 km), in accordance with the International Ski Federation regulations (Figure 1). During the race, each Para and AB XC skier was continuously tracked with a Catapult device (OptimEye S5, Catapult Innovations, Melbourne, Australia) with integrated 10 Hz global navigation satellite system (GNSS) receiver and an inertial measurement unit (IMU) providing 100 Hz triaxial accelerometer and gyroscope data, positioned in a tight fitting-vest on the skier's upper back under the race bib. All XC skiers raced on the same day. The time-trials started in the morning, with the female AB XC skiers racing first followed by the male AB XC skiers. Thereafter, the female and male Para XC skiers completed their race within the same time range. The start interval between each Para and AB XC skier was 30–60 s. Every athlete used their own ski equipment, including skis, poles, boots, and ski base material (including grinds, structure, and waxing), with adjustments being made by each team's waxing crew according to individual preferences and daily conditions. The weather conditions were stable throughout the day, with a snow temperature of −12 °C and an air temperature between −4 to −7 °C during all the races. The snow friction was measured as 0.023 in the middle of the day. The course was covered with hard-packed snow and machine-prepared directly before the races of the AB and Para XC skiers.

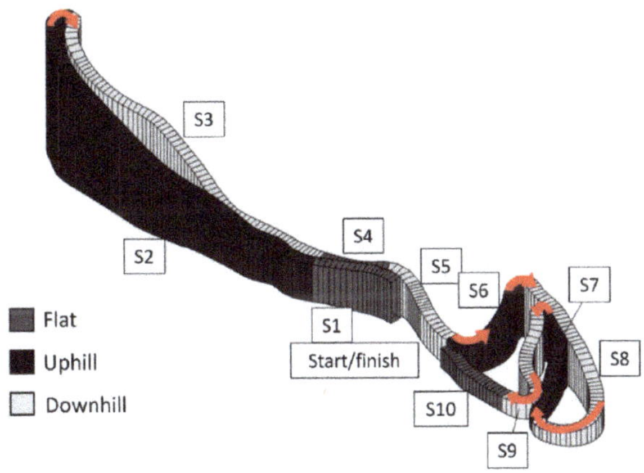

Figure 1. The 2.5 km XC skiing race course divided into the 10 segments according to the elevation difference, including three uphill, three flat, and four downhill segments. Six turns were distributed over the 2.5 km lap (red arrow). With different placement of the start and finish, there is a gap in the 2.5 km course, which was removed from the analyses. (S) Segment (length (meter), incline range (%)); S1: 131 m, −2.6–1.6%; S2: 543 m, 1.7–12.4%; S3: 509 m, −11.7–0.2%; S4 100 m, 0.7–2.8%; S5: 156 m, −6.0–0.0%; S6: 166 m, 1.3–12.7%; S7: 339 m, −8.5–−0.4%; S8: 200 m, 1.2–16%; S9: 183 m, −10.1–−1.0%; S10: 138 m, 0.0–1.6%.

2.3. Measurements

Time, positioning, altitude, and movement data for all Para and AB XC skiers were measured continuously during the race with the Catapult devices. The speed data were derived from time differentiation of the position data. Prior to the data collection, the Catapult devices were placed outside in an open space for a minimum of 10 min to ensure GNSS lock and allow acquisition of satellite signals. Recently, Gløersen et al. [22] have validated the Catapult devices for position, speed, and time analyses in AB XC skiing against a geodetic, multi-frequency receiver, with a horizontal plane position error of 1.04 m (third quartile, Q3), horizontal plane speed of 0.072 $m \cdot s^{-1}$ (IQR), and time precision between 0.13–0.36 s. Similar accuracy is expected when using the Catapult devices for analysis of performance as done in the current study.

2.4. Data Analysis

The length and elevation profile of the race course were obtained from the GNSS data measured with the Catapult devices that were used to track the Para and AB XC skiers. Based on the positioning and altitude data, the course was divided into segments consisting of either uphill, flat, and downhill terrain. Each segment began and ended with an evident change in the gradient of the course. The uphill and downhill segments were characterized by a minimum elevation difference of 4 m from the beginning to the end of the segment. Undulating terrain with a smaller elevation difference between adjacent uphill, flat, and downhill segments were merged into one single flat segment. Overall, the 2.5 km course was divided into 10 segments, with three uphill, three flat, and four downhill segments that made up 37%, 15%, and 48% of the course, respectively. Additionally, the 2.5 km course included six turns (Figure 2). Different placements of the start and finish of the race resulted in a gap in the 2.5 km course, which was removed from the analyses. The actual distance covered for each segment was calculated using the elevation difference from the beginning to the end of the segment and the horizontal length of the segment.

Data of speed and time were interpolated by distance for each lap for both Para and AB XC skiers. Further, the average speed and time over the four or six laps were calculated and used in the analyses. In order to compare Para and AB XC skiers, average values of speed and time were calculated for each group of female and male AB XC skiers. Furthermore, the continuous speed and time differences between the male Para and male AB XC skiers and between the female Para and female AB XC skiers were calculated. The proportion of time in the different terrains was calculated for each Para XC skier, the female, and the male AB XC skiers (mean ± standard deviation (SD)).

From the movement data of Para and AB XC skiers, measured by IMUs in the Catapult devices, automatic sub-technique classification was done by employing a K-Nearest Neighbour algorithm while using a 2 s sliding window approach (200 samples) with 95% overlap [15]. The classifier uses the low-pass filtered z-components of the accelerometer and gyroscope data as input, with the z-axis defined in the frontal direction of the participant. The same classifier was used for all skiers. The classifier was validated on AB XC skiers with a per-distance classification accuracy of 96% [6,21,23]. To apply the framework and accompanying algorithms to Para XC skiers with similar accuracy, visual examination of the classification was conducted by comparing the graphical representation of filtered accelerometer and gyroscope signals with examples that typically represent the various sub-techniques. Thereby, around 10% of the cycles from the automatic classification were manually corrected. The sub-techniques were classified as DP, DK, DIA (including both DIA and HRB), and OTHER. OTHER primarily included the tuck position, but also turn techniques and cycles that did not fulfill the above-specified criteria. At higher speeds (i.e., 7 to 10 $m \cdot s^{-1}$), OTHER almost solely contained the tuck position.

2.5. Statistical Analysis

In this case-series, descriptive comparison was made for speed, time, and sub-technique distribution between the Para and AB XC skiers, exemplifying the feasibility of the framework based on micro-sensor technology employed in the field for Para XC skiing. Data processing and calculations were done using MATLAB R2018a (version 9.7.0.1190202, MathWorks, Natick, MA) and Microsoft Office Excel 2016 (Microsoft Corporation, Redmond, WA, USA).

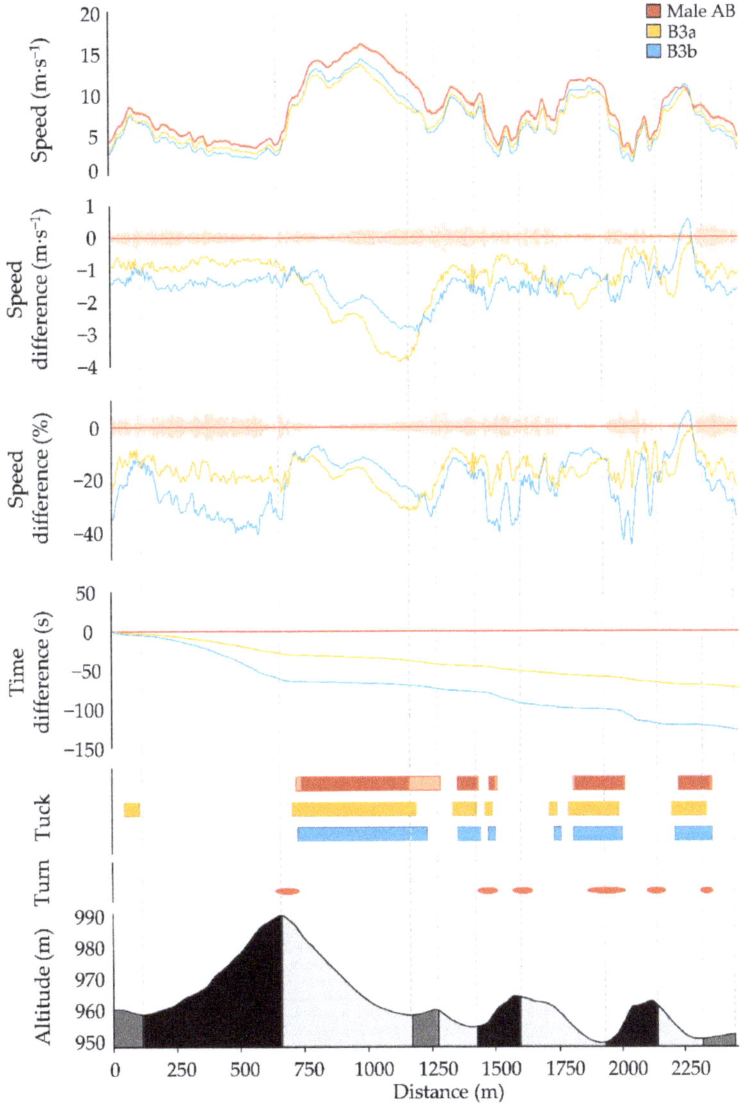

Figure 2. Comparison of average male AB XC skiers (dark red; standard deviation light pink), male B3a XC skier (yellow), and male B3b XC skier (blue) for the six laps during the race with respect to average speed, absolute speed difference, relative speed difference, accumulated time difference, and tuck. Course details are visualized in the lower part of the figure; turns (red dashes) and altitude profile of the 2.5 km race course, with uphill (black), flat (dark gray), and downhill (light gray) segments.

3. Results

3.1. Performance

The general speed fluctuation patterns of Para and AB XC skiers were similar throughout the race, although Para XC skiers consistently competed at slower speed (Figures 2 and 3, Table 2). Accordingly, the female LW4 skier was 4:26 min slower compared to the female AB XC skiers across the entire race, whereas the male B3a skier was 7:21 min slower and the male B3b skier 12:47 min slower compared to the male AB XC skiers. Compared to male/female AB skiers, male/female Para skiers displayed 19/14% slower average speed with the largest time loss (65 ± 36/35 ± 6 s/lap) found in uphill terrain (Table 2). The relative speed difference between the Para and AB XC skiers was highest in uphill and flat terrain, followed by downhill terrain (Table 2).

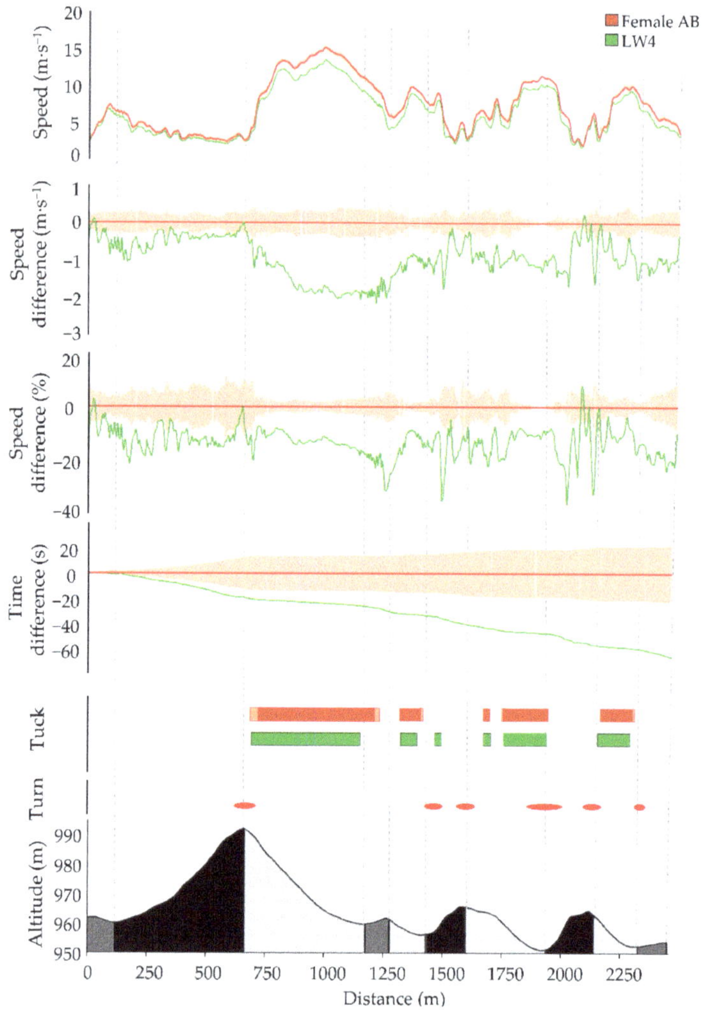

Figure 3. Comparison of average female AB XC skiers (dark red; standard deviation light pink) and female LW4 XC skier (green) for the four laps during the race with respect to average speed, absolute speed difference, relative speed difference, accumulated time difference, and tuck. Course details are visualized in the lower part of the figure; turns (red dashes) and altitude profile of the 2.5 km race course, with uphill (black), flat (dark gray), and downhill (light gray) segments.

Table 2. Proportion of skiing time in different terrain (%), absolute average speed (m·s^{-1}), relative speed difference (% of AB XC skiers), and time loss relative to AB XC skiers of same sex per lap (s) for the Para and AB XC skiers.

		LW4	Female AB	B3a	B3b	Male AB
Proportion of time in different terrain (%)	Uphill	57	58 ± 16	54	59	54 ± 14
	Flat	1.5	14 ± 1	14	14	14 ± 1
	Downhill	28	28 ± 3	31	27	32 ± 3
Absolute average speed (m·s^{-1})	Overall	6.0 ± 2.3	6.9 ± 2.6	6.7 ±	6.4 ± 2.4	7.9 ± 2.6
	Uphill	3.7 ± 0.5	4.3 ± 0.6	4.6 ± 0.6	3.9 ± 0.5	5.3 ± 0.6
	Flat	5.1 ± 0.7	6.2 ± 1.2	6.0 ± 0.6	5.7 ± 0.8	7.5 ± 1.4
	Downhill	8.3 ± 1.6	9.5 ± 1.9	8.7 ± 1.2	8.9 ± 1.5	10.3 ± 1.9
Relative speed difference (% of AB XC skiers)	Overall	14 ± 4	100	16 ± 5	20 ± 7	100
	Uphill	14 ± 2	100	14 ± 2	26 ± 3	100
	Flat	16 ± 8	100	19 ± 7	24 ± 4	100
	Downhill	12 ± 1	100	15 ± 4	14 ± 5	100
Time loss per lap (s)	Overall	65 ± 10		72 ± 7	126 ± 17	
	Uphill	35 ± 6		13 ± 11	30 ± 24	
	Flat	11 ± 3		4 ± 1	5 ± 1	
	Downhill	19 ± 2		6 ± 4	5 ± 3	

3.2. Sub-Technique Distribution

The female Para/AB XC skiers utilized on average DP, DK, DIA, and OTHER, 61/43%, 15/10%, 25/47%, and 0/0% of the distance at lower speed ranges (i.e., 2.75 to 4.75 m·s^{-1}), respectively, while the corresponding numbers for male Para/AB XC skiers were 58/18%, 1/13%, 40/69%, and 1/0%. The female Para/AB XC skiers utilized on average DP and OTHER (i.e., tuck position), 26/52% and 74/48% of the distance at higher speed ranges (i.e., 7 to 10 m·s^{-1}), respectively, while the corresponding numbers for male Para/AB XC skiers were 29/66% and 71/34% (Figures 4 and 5).

The male B3 and female LW4 XC skiers used the tuck position at similar positions during the race course as the male and female AB XC skiers (Figures 2 and 3) but employed them at a slower speed compared to the AB XC skiers (Figures 4 and 5).

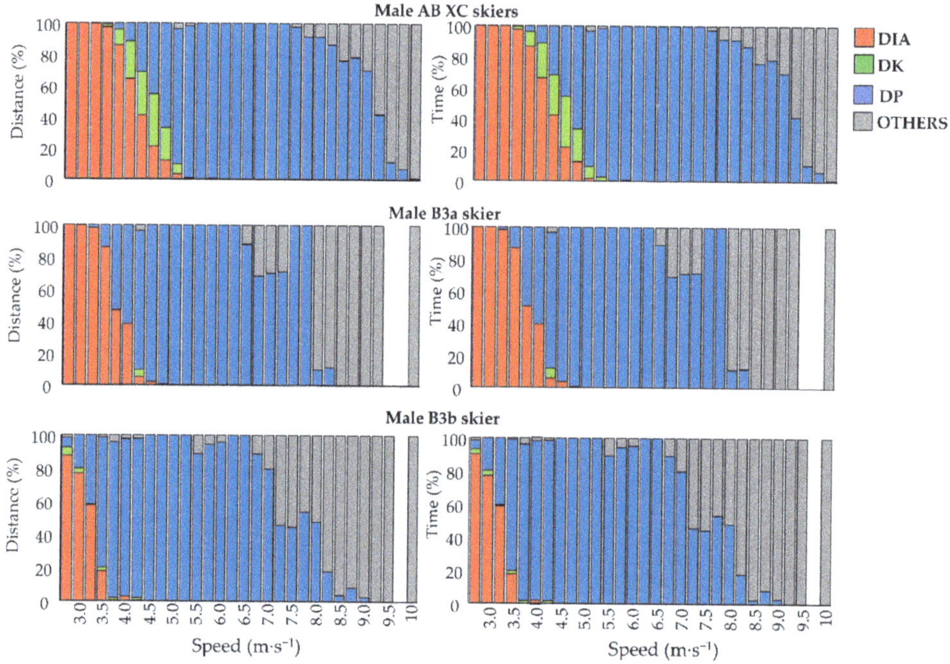

Figure 4. Distribution of sub-techniques over different speed-intervals for male AB, the B3a, and the B3b XC skiers per distance and time. Diagonal stride (DIA; red); Kick double poling (DK; green); Double poling (DP; blue); Tuck position and turn technique (OTHER; gray); White sections illustrate that the skier did not use these speeds.

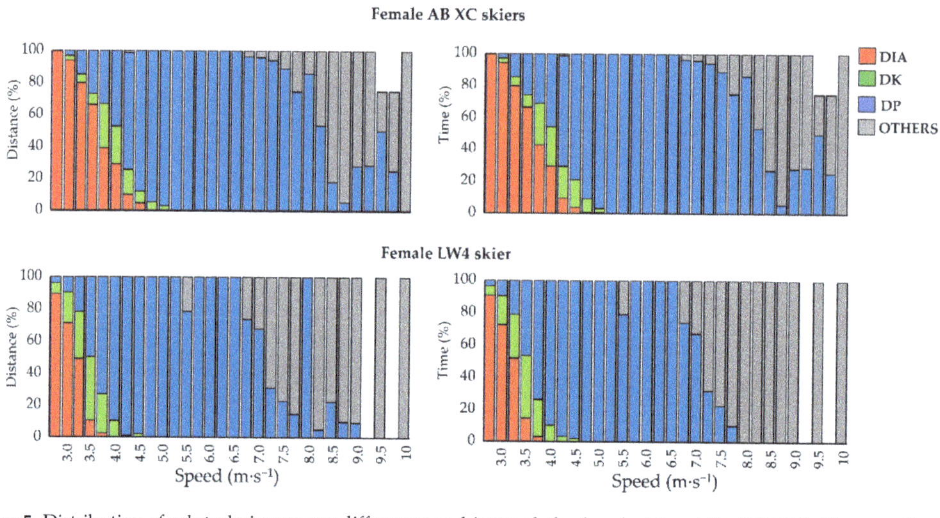

Figure 5. Distribution of sub-techniques over different speed-intervals for female AB and the LW4 XC skiers per distance and time. Diagonal stride (DIA; red); Kick double poling (DK; green); Double poling (DP; blue); Tuck position and turn technique (OTHER; gray). Blank sections for AB XC skiers and the LW4 XC skier illustrate that one of the skiers didn't used these speeds.

4. Discussion

The framework based on micro-sensor technology allowed us to descriptively compare performance and sub-technique selection between standing Para and AB XC skiers in a classical XC skiing race. Using case-series, we revealed that male/female Para skiers displayed 19/14% slower average speed with the largest time loss (65 ± 36/35 ± 6 s per lap) in uphill terrain compared to their AB counterparts. Furthermore, the Para XC skiers utilized a larger proportion of DP than DIA and DK per distance at low speeds (i.e., 2.75 to 4.75 $m{\cdot}s^{-1}$) and a larger proportion of tuck than DP per distance at high speeds (i.e., 7 to 10 $m{\cdot}s^{-1}$). Since we are able to distinguish clear differences between Para and AB XC skiers, we propose that the framework is feasible for future use in large-scale investigations of performance at international competitions. Especially, this is the case for more in-depth investigations on the effect of terrain and external conditions on the time factor across Para XC skiing classes.

In line with both the large speed difference and the amount of skiing time spent in uphill terrain (women: LW4: 57% and AB: 58 ± 16%; men: B3: 57 ± 3% and AB: 54 ± 14%), the time loss between the Para and AB XC skiers of the same sex were largest in this terrain. This is in accordance with previous studies in AB XC skiers who spent ~50% of skiing time in uphill terrain, in which the largest performance differences were found [4,19,24,25]. Interestingly, the relative speed difference between Para and AB XC skiers of the same sex in flat terrain was relatively similar to the difference found in uphill terrain. This differs from AB XC skiing, where the relative speed difference is less in flat compared to uphill terrain among different level skiers [19,24,25]. The large relative speed difference in flat terrain may be caused by a reduced balance and motor control of both the female LW4 [26] and the male B3 [26,27] XC skiers due to their impairments. This could have impacted the movement patterns on flat terrain at high speeds.

This is the first study to perform automatic sub-technique classification in standing Para XC skiers during an entire XC skiing race. The comparison to AB XC skiers revealed that, Para XC skiers utilized different proportions of the various sub-techniques at given low and high absolute speeds. This differs from a previous comparison between female and male AB XC skiers during a classical 10 km XC skiing race, that revealed similar proportions of used sub-techniques at same absolute speeds, despite a slower average speed employed by the female AB XC skiers [15]. Regarding the different sub-techniques between Para and AB XC skiers found here, this may be caused by the different (and possibly less technically demanding) coordination stability between arms and legs in DP compared to DIA and DK [28]. Accordingly, the greater use of DP than DIA by the Para XC skiers compared to AB XC skiers at low speeds could be speculated to be caused by the fact that DP is more suitable than DIA and DK for the Para XC skiers in the investigated speeds. In addition, the leg thrust time (i.e., the time during which the ski is in contact with the ground in a leg stride in DIA and DK) in AB XC skiing is suggested to have a speed limit that triggers the transition from DIA to DP [12,29]. For Para XC skiers such limits may be present at a lower speed. Furthermore, the Para XC skiers used a larger proportion of the distance in tuck position than DP at high speeds. Altogether, this indicates that different speed thresholds are present for the choice of classical sub-techniques in Para XC skiers than those suggested by research on AB XC skiers [6,15,17,18]. While this is likely due to disability-related limitations in the Para XC skiers, this still needs further investigation with larger sample sizes during international competitions.

Even though the proposed framework seems feasible for investigating performance also in Para XC skiers, it has some methodological limitations. The framework has only been tested in one race course and under the given external conditions. Since different external conditions (e.g., race courses with and without trees or other obstacles, weather conditions, low- vs. high-speed race courses, etc.) can affect the accuracy of the GNSS receiver, athletes who train and compete in different environments should take this into account and future studies should test the feasibility of the framework under different external conditions. Furthermore, the framework used for automatic sub-technique classification should, in future studies, be adapted for all sitting and standing Para XC skiing categories and validated in larger populations. In line with this, the framework was only applied among standing physically and visually impaired Para XC skiers in the current study, and can, therefore, only be regarded feasible in these athletes.

Practical Applications

The framework can be used to provide information on where and why Para XC skiers lose or win time compared to their competitors in a race course, as well as the effect of terrain and external conditions on the time factors used in the classification process of Para XC skiers. This could help Para XC skiers to individually define targeted training and competition strategies. In addition, our approach can be used on larger groups of Para XC skiers to provide a more detailed understanding on the influence of sub-technique and terrain on the differences between disabilities, categories, and sexes.

Furthermore, the sub-technique analyses provide information on the specific speeds and terrains where Para XC skiers employ the different sub-techniques, as well as how corresponding temporal patterns within these sub-techniques influence performance. In this study, we found a different distribution of the classical sub-techniques between the standing Para and AB XC skiers both at low and high speeds during an XC skiing race. Together with the large relative speed difference in flat terrain with high skiing speed, the movement pattern of the Para XC skiers seems to be differently exposed to the high speed than AB XC skiers, hence affecting the selected sub-technique. Such information is useful for athletes and coaches when deciding what type of training the different skiers should prioritize (e.g., improvement of technique execution and balance at high skiing speed, or development of aerobic capacity to increase performance in uphill terrain).

5. Conclusions

This study evaluated the feasibility of a framework for analyses of performance and sub-technique selection in a heterogenous group of Para XC skiers with different disabilities during a classical XC skiing race. A descriptive comparison of performance and sub-technique selection between Para and AB XC skiers indicated that the largest time loss between the Para and AB XC skiers was found in the uphill terrain. In contrast to the larger speed differences normally found in uphill terrain between performance-levels or sexes within AB skiers, the Para XC skiers displayed a similar relative speed difference compared to AB skiers in flat as in uphill terrain. This may be caused by a reduced balance and motor control of the Para XC skiers due to their impairments and also impact the movement pattern on flat terrain at high speeds. Furthermore, the Para XC skiers more frequently selected DP than DIA and DK at low speeds. Speculatively, DP could be more suitable than DIA and DK due to its lower coordinative demands for Para XC skiers who struggle with stability/coordination. Additionally, the Para XC skiers used a larger proportion of the distance in tuck position than DP at high speeds. Notably, this indicates different speed thresholds of the classical sub-techniques for Para XC skiers compared to AB XC skiers. Altogether, we hypothesize that disability impacts the selection of sub-technique among standing Para XC skiers, which could be examined by using the framework in large-scaled international investigations. Additionally, the framework opens up the possibility to investigate the effect of terrain and external conditions on the time factor across Para XC skiing classes.

Author Contributions: Conceptualization, C.H.C., J.K.B., J.K. and Ø.S.; Methodology, E.M.B.P., J.K.B. and Ø.S.; Software, J.K.; Formal Analysis, J.K., E.M.B.P. and C.H.C.; Investigation, J.K.B. and P.H.; Writing—Original Draft Preparation, C.H.C., J.K.B. and Ø.S.; Writing—Review and Editing, P.H., E.M.B.P. and J.K.; Visualization, C.H.C. and J.K. All authors have read and agreed to the published version of the manuscript.

Funding: The study was funded by the Centre for Elite Sports Research, Dept of Neuromedicine and Movement Science, Norwegian University of Science and Technology, Trondheim, Norway.

Institutional Review Board Statement: The study was approved by the Norwegian Centre for Research Data (ID 49865/3/IJJ) and conducted in line with the Declaration of Helsinki.

Informed Consent Statement: Informed consent was obtained from all subjects involved in the study.

Acknowledgments: The authors would like to thank the Para and able-bodied XC skiers for their participation in the study.

Conflicts of Interest: The authors declare no conflict of interest.

References

1. IPC. *Explanatory Guide to Paralympic Classification—Paralympic Winter Sports*; World Para Nordic Skiing: Bonn, Germany, 2016.
2. The International Paralympic Committee, World Para Snow Sports. Available online: https://www.paralympic.org/nordic-skiing/about (accessed on 24 February 2021).
3. IPC. *World Para Nordic Skiing Rules and Regulations Cross-Country Skiing and Biathlon*; World Para Nordic Skiing: Bonn, Germany, 2018.
4. Andersson, E.; Supej, M.; Sandbakk, Ø.; Sperlich, B.; Stöggl, T.; Holmberg, H.C. Analysis of sprint cross-country skiing using a differential global navigation satellite system. *Eur. J. Appl. Physiol.* **2010**, *110*, 585–595. [CrossRef] [PubMed]
5. Losnegard, T. Energy system contribution during competitive cross-country skiing. *Eur. J. Appl. Physiol.* **2019**, *119*, 1675–1690. [CrossRef]
6. Solli, G.S.; Kocbach, J.; Seeberg, T.M.; Tjønnås, J.; Rindal, O.M.; Haugnes, P.; Torvik, P.Ø.; Sandbakk, Ø. Sex-based differences in speed, sub-technique selection, and kinematic patterns during low- and high-intensity training for classical cross-country skiing. *PLoS ONE* **2018**, *13*, e0207195. [CrossRef]
7. Welde, B.; Stöggl, T.L.; Mathisen, G.E.; Supej, M.; Zoppirolli, C.; Winther, A.K.; Pellegrini, B.; Holmberg, H.C. The pacing strategy and technique of male cross-country skiers with different levels of performance during a 15-km classical race. *PLoS ONE* **2017**, *12*, e0187111. [CrossRef]
8. Holmberg, H.C.; Lindinger, S.; Stöggl, T.; Eitzlmair, E.; Müller, E. Biomechanical analysis of double poling in elite cross-country skiers. *Med. Sci. Sports Exerc.* **2005**, *37*, 807–818. [CrossRef]
9. Stöggl, T.L.; Holmberg, H.C. Double-Poling Biomechanics of Elite Cross-country Skiers: Flat versus Uphill Terrain. *Med. Sci. Sports Exerc.* **2016**, *48*, 1580–1589. [CrossRef]
10. Göpfert, C.; Holmberg, H.C.; Stöggl, T.; Müller, E.; Lindinger, S.J. Biomechanical characteristics and speed adaptation during kick double poling on roller skis in elite cross-country skiers. *Sports Biomech.* **2013**, *12*, 154–174. [CrossRef] [PubMed]
11. Dahl, C.; Sandbakk, Ø.; Danielsen, J.; Ettema, G. The Role of Power Fluctuations in the Preference of Diagonal vs. Double Poling Sub-Technique at Different Incline-Speed Combinations in Elite Cross-Country Skiers. *Front. Physiol.* **2017**, *8*, 94. [CrossRef]
12. Pellegrini, B.; Zoppirolli, C.; Bortolan, L.; Holmberg, H.C.; Zamparo, P.; Schena, F. Biomechanical and energetic determinants of technique selection in classical cross-country skiing. *Hum. Mov. Sci.* **2013**, *32*, 1415–1429. [CrossRef]
13. Andersson, E.; Stöggl, T.; Pellegrini, B.; Sandbakk, Ø.; Ettema, G.; Holmberg, H.C. Biomechanical analysis of the herringbone technique as employed by elite cross-country skiers. *Scand. J. Med. Sci. Sports* **2014**, *24*, 542–552. [CrossRef] [PubMed]
14. Bucher Sandbakk, S.; Supej, M.; Sandbakk, Ø.; Holmberg, H.C. Downhill turn techniques and associated physical characteristics in cross-country skiers. *Scand. J. Med. Sci. Sports* **2014**, *24*, 708–716. [CrossRef] [PubMed]
15. Strøm Solli, G.; Kocbach, J.; Bucher Sandbakk, S.; Haugnes, P.; Losnegard, T.; Sandbakk, Ø. Sex-based differences in sub-technique selection during an international classical cross-country skiing competition. *PLoS ONE* **2020**, *15*, e0239862. [CrossRef]
16. Ettema, G.; Kveli, E.; Øksnes, M.; Sandbakk, Ø. The role of speed and incline in the spontaneous choice of technique in classical roller-skiing. *Hum. Mov. Sci.* **2017**, *55*, 100–107. [CrossRef]
17. Marsland, F.; Anson, J.; Waddington, G.; Holmberg, H.C.; Chapman, D.W. Macro-Kinematic Differences Between Sprint and Distance Cross-Country Skiing Competitions Using the Classical Technique. *Front. Physiol.* **2018**, *9*, 570. [CrossRef] [PubMed]
18. Marsland, F.; Mackintosh, C.; Holmberg, H.C.; Anson, J.; Waddington, G.; Lyons, K.; Chapman, D. Full course macro-kinematic analysis of a 10 km classical cross-country skiing competition. *PLoS ONE* **2017**, *12*, e0182262. [CrossRef]
19. Sandbakk, Ø.; Losnegard, T.; Skattebo, Ø.; Hegge, A.M.; Tønnessen, E.; Kocbach, J. Analysis of Classical Time-Trial Performance and Technique-Specific Physiological Determinants in Elite Female Cross-Country Skiers. *Front. Physiol.* **2016**, *7*, 326. [CrossRef]
20. Sollie, O.; Gløersen, Ø.; Gilgien, M.; Losnegard, T. Differences in pacing pattern and sub-technique selection between young and adult competitive cross-country skiers. *Scand. J. Med. Sci. Sports* **2021**, *31*, 553–563. [CrossRef]
21. Seeberg, T.M.; Tjønnås, J.; Rindal, O.M.; Haugnes, P.; Dalgard, S.; Sandbakk, Ø. A multi-sensor system for automatic analysis of classical cross-country skiing techniques. *Sports Eng.* **2017**, *20*, 313–327. [CrossRef]
22. Gløersen, Ø.; Kocbach, J.; Gilgien, M. Tracking Performance in Endurance Racing Sports: Evaluation of the Accuracy Offered by Three Commercial GNSS Receivers Aimed at the Sports Market. *Front. Physiol.* **2018**, *9*, 1425. [CrossRef]
23. Rindal, O.M.; Seeberg, T.M.; Tjønnås, J.; Haugnes, P.; Sandbakk, Ø. Automatic Classification of Sub-Techniques in Classical Cross-Country Skiing Using a Machine Learning Algorithm on Micro-Sensor Data. *Sensors* **2017**, *18*, 75. [CrossRef] [PubMed]
24. Bolger, C.M.; Kocbach, J.; Hegge, A.M.; Sandbakk, Ø. Speed and heart-rate profiles in skating and classical cross-country skiing competitions. *Int. J. Sports Physiol. Perform* **2015**, *10*, 873–880. [CrossRef] [PubMed]

25. Sandbakk, Ø.; Ettema, G.; Leirdal, S.; Jakobsen, V.; Holmberg, H.C. Analysis of a sprint ski race and associated laboratory determinants of world-class performance. *Eur. J. Appl. Physiol.* **2011**, *111*, 947–957. [CrossRef] [PubMed]
26. Norges Skiforbund. Available online: https://www.skiforbundet.no/langrenn (accessed on 12 February 2021).
27. Klavina, A.; Jekabsone, I. Static Balance of Persons with Intellectual Disabilities, Visual Impairments and without Disabilities. *Eur. J. Adapt. Phys. Act.* **2014**, *7*, 50–57. [CrossRef]
28. Cignetti, F.; Schena, F.; Zanone, P.G.; Rouard, A. Dynamics of coordination in cross-country skiing. *Hum. Mov. Sci.* **2009**, *28*, 204–217. [CrossRef]
29. Nilsson, J.; Tveit, P.; Eikrehagen, O. Effects of Speed on Temporal Patterns in Classical Style and Freestyle Cross-Country Skiing. *Sports Biomech.* **2004**, *3*, 85–107. [CrossRef]

sensors

MDPI

Article

A Novel Sensor Foil to Measure Ski Deflections: Development and Validation of a Curvature Model

Christoph Thorwartl [1,*], Josef Kröll [1], Andreas Tschepp [2], Philipp Schäffner [2], Helmut Holzer [3] and Thomas Stöggl [1,4]

1 Department of Sport and Exercise Science, University of Salzburg, Schlossallee 49, 5400 Hallein/Rif, Austria; josef.kroell@sbg.ac.at (J.K.); thomas.stoeggl@sbg.ac.at (T.S.)
2 Joanneum Research Forschungsgesellschaft mbH, Franz-Pichler-Straße 30, 8160 Weiz, Austria; Andreas.Tschepp@joanneum.at (A.T.); Philipp.Schaeffner@joanneum.at (P.S.)
3 Atomic Austria GmbH, Atomic Strasse 1, 5541 Altenmarkt, Austria; Helmut.Holzer@atomic.com
4 Athlete Performance Center, Red Bull Sports, Brunnbachweg 71, 5303 Thalgau, Austria
* Correspondence: christoph.thorwartl@sbg.ac.at; Tel.: +43-664-3455681

Abstract: The ski deflection with the associated temporal and segmental curvature variation can be considered as a performance-relevant factor in alpine skiing. Although some work on recording ski deflection is available, the segmental curvature among the ski and temporal aspects have not yet been made an object of observation. Therefore, the goal of this study was to develop a novel ski demonstrator and to conceptualize and validate an empirical curvature model. Twenty-four PyzoFlex® technology-based sensor foils were attached to the upper surface of an alpine ski. A self-developed instrument simultaneously measuring sixteen sensors was used as a data acquisition device. After calibration with a standardized bending test, using an empirical curvature model, the sensors were applied to analyze the segmental curvature characteristic (m^{-1}) of the ski in a quasi-static bending situation at five different load levels between 100 N and 230 N. The derived curvature data were compared with values obtained from a high-precision laser measurement system. For the reliability assessment, successive pairs of trials were evaluated at different load levels by calculating the change in mean (CIM), the coefficient of variation (CV) and the intraclass correlation coefficient (ICC 3.1) with a 95% confidence interval. A high reliability of CIM −1.41–0.50%, max CV 1.45%, and ICC 3.1 > 0.961 was found for the different load levels. Additionally, the criterion validity based on the Pearson correlation coefficient was $R^2 = 0.993$ and the limits of agreement, expressed by the accuracy (systematic bias) and the precision (SD), was between $+9.45 \times 10^{-3}$ m^{-1} and -6.78×10^{-3} m^{-1} for all load levels. The new measuring system offers both good accuracy (1.33×10^{-3} m^{-1}) and high precision (4.14×10^{-3} m^{-1}). However, the results are based on quasi-static ski deformations, which means that a transfer into the field is only allowed to a limited extent since the scope of the curvature model has not yet been definitely determined. The high laboratory-related reliability and validity of our novel ski prototype featuring PyzoFlex® technology make it a potential candidate for on-snow application such as smart skiing equipment.

Keywords: bending sensors; flexion; PyzoFlex; ski bending; ski deflection

Citation: Thorwartl, C.; Kröll, J.; Tschepp, A.; Schäffner, P.; Holzer, H.; Stöggl, T. A Novel Sensor Foil to Measure Ski Deflections: Development and Validation of a Curvature Model. *Sensors* **2021**, *21*, 4848. https://doi.org/10.3390/s21144848

Academic Editor: Giuseppe Vannozzi

Received: 1 June 2021
Accepted: 14 July 2021
Published: 16 July 2021

1. Introduction

1.1. Ski Deflection in Alpine Skiing

With the development of shaped skis in the late 1990s, performing carved turns has become possible. A carved turn is defined by minimal or no lateral ski displacement relative to the track, and therefore, each point along the ski edge follows the path of the proceeding one [1–4]. In contrast, during a skidding turn, a point along the edge does not follow the path of the proceeding one but slides sideways along the slope [1–4]. Skidded and carved turns differ not only in terms of ski trajectory but also regarding ski deflection. While in carved turns, ski deflection is a prerequisite to meet the above definition, in skidded turns, the ski does not necessarily have to be deflected. For practical reasons, the skiing motion is often described either as skidding or carving, but carving and skidding are not dichotomous processes; rather, both can occur at the same time in different segments of the ski's length [3]. To represent the inhomogeneous deflection progression, it would thus be an oversimplification to fit a circle with a constant radius into the deflection line of the ski. One would need several circles with different radius instead. In addition to the segmental differences in deflection along the ski, the temporal sequence of deflection varies in short time periods within a turning phase [5]. However, temporal and segmental changes in deflection during alpine skiing give an essential insight into how the ski-snow interaction proceeds and provides information about the quality of a turn.

1.2. Status Quo and Limitations of Research in Measuring Ski Deflection

Standardized laboratory measurements are suitable for a more detailed examination of ski deflection. One of the most often applied methodologies to deform a ski under standardized conditions is the application of a 3-point bending test. To quantify the deflection profile of the ski, the vertical displacement (w) [6–10], the angular deformation [11] or the curvature [12] are typically measured. Most applications measure the vertical deflection by using displacement transducers [6,7], linear potentiometers [8], laser transducers [9] or linear variable differential transformers [10]. Additionally, motion capture systems are used to detect w [10]. Less common is the angular deformation measurement with an optical encoder [11] or the curvature detection with a digital radius gauge, whereby the last was conducted for snowboards [12]. From a mathematical perspective, the angle (gradient) corresponds to the first derivative (dw/dx) and the curvature to the second derivative $\left(d^2w/dx^2\right)$ of the deflection line. Therefore, these physical quantities can be transformed into each other under consideration of the boundary conditions. A detailed consideration or analysis of the angular or curvature progression has not been carried out so far, rather it was used implicitly to calculate the bending stiffness progression, which is a fundamental design parameter of alpine skis [6,7,9–13]. To operationalize the segmental ski deflection, $w(x)$ is not sufficient because only a qualitative description would be possible. To determine the segment where the largest deflection occurs in a quantitative way, either the angle or the curvature is needed. In summary, there are a number of studies that have dealt with standardized ski deflection in the laboratory, but the exact deflection characteristic along the ski was not analyzed in detail.

The predominant sensors that have been used up to now to determine the ski's deflection during alpine skiing are strain gauges [5,14–19]. In several articles, an aluminum deflection sensor beam with integrated strain cells was applied to the upper surface of the ski [5,15–17]. A mathematical method was used to model the deflection of the ski's running surface based on measured bending and torsion [17]. The snow measurement showed that the deflection consists predominantly of bending and that the torsional deformation has only a minor influence on the shape of the running surface [17]. It was indicated that the ski deflection varied within a few milliseconds [5], and the inner ski was less bent than the outer ski [17]. The construction of the ski with the aluminum deflection sensor beam is very complex, which affects the dynamic properties of the ski [17] and makes a series implementation not possible. Fauve et al. [19] analyzed the influence of snow penetration strength and loading on ski deflection and Schindelwig et al. [18] the influence of bending and torsional stiffness with regard to ski deflection. In another study, the captured turn radii from a differential global navigation satellite system were compared with the calculated radii from strain gauges [14]. A smart ski prototype with a feedback system was also equipped with bending sensors, although the signal was not calibrated and thus only has relative validity [20].

Research about ski deflection, in general, is scarce and, more specifically, the analysis of ski deflection in view of local and temporal variations. Yoneyama et al. [17] differentiated between front and rear ski segments by calculating dw/dx in relation to the boot center coordinate system. However, the gradient was not considered further in the analysis. Other authors either described the segmental deflection of the ski running surface qualitatively with $w(x)$ [5,16,19] or calculated the overall bending radius ($R = \overline{1/(d^2w/dx^2)}$) along the running surface [14]. The temporal deflection was also considered only at a rudimentary level. So far, only proof of concept studies with superficial validation approaches was provided for the strain gauge-based prototypes, but no concrete applications or commercial spin-offs took place. This may be related to the complex serial implementation of strain gauge sensors. From a scientific point of view, the strain gauge-based systems were not even tested for reliability or validity. In summary, some promising approaches to quantify the ski deflection during alpine skiing were pursued [5,14–20], but there is a paucity of approaches regarding segmental and temporal deflection with respect to scientific evaluation. Consequently, no useful real-world application is currently available.

1.3. Novel Prototype for Ski Deflection Detection

Here, we present a new approach for ski deflection measurement relying on PyzoFlex® technology (www.pyzoflex.com accessed on 15 July 2021, Joanneum Research Forschungsgesellschaft m.b.H, Franz-Pichler Str. 30, 8160 Weiz, Austria). The novel measurement system will be used to resolve the segmental curvature along the ski and obtain a more differentiated picture of the ski deflection. The PyzoFlex® foils can be produced by a cost-effective screen-printing process, and therefore, the sensors are commercially highly relevant. Different arrangements, sizes and shapes of the sensors can be specified and easily printed. Furthermore, the foils can be readily applied to skis with the help of double-sided adhesive tape. The ski equipment and the dynamic behavior of the skis are hardly influenced by the foils since the sensors are flexible and practically massless. First, rudimentary on-snow proof of concept tests with a simple sensor layout served plausible raw signals. Therefore, the PyzoFlex® technology seems to have application potential in the context of alpine skiing.

1.4. Goals and Research Questions

The aims of this study were to (a) develop a ski prototype with PyzoFlex® technology-based sensor foils, (b) establish a curvature model based on the PyzoFlex® signal applying standardized bending conditions and (c) test the reliability and validity of this model.

2. Materials and Methods

2.1. Development of the Ski Demonstrator

The PyzoFlex® sensors are based on a ferroelectric polymer material and are thus intrinsically piezoelectric and pyroelectric. The sensor foil is constructed in a sandwich structure. The basis of the sensor foil is a transparent polyethylene terephthalate substrate (PET), which serves as a carrier for the printed ferroelectric material. The piezoelectric and pyroelectric sensor material Poly(vinylidene fluoride–trifluoroethylene), denoted as P(VDF-TrFE), is located between the top and bottom electrodes [21]. When a stimulus is applied to the sensor, the surface charge density of the sensor material changes, inducing a change of compensation charges in the electrodes.

To determine the ski deflection, an arrangement of 24 single sensor elements within a foil was first considered and designed in a computer-aided design program. The production of the sensor foils with this design was achieved with an industrially screen-printing process with a carrier substrate (PET) of the sensors with 125 µm and the printed layers (base electrode, active material, cover electrode, silver tracks and protective layer) with a total of about 20 µm. The maximum film size to be produced is limited to A3 format due to the available screen-printing equipment. In the case of the ski design, this resulted in three film elements, which were distributed separately over the ski (Figure 1a). The separation of the three sensor elements—2 foil pieces with 9 sensors, 1 foil piece with 6 sensors—was achieved with the use of a Trotec laser cutter. The electrical connections (printed silver lines) were chosen in such a way that the 24 sensors can be wired via three interfaces. The interface between sensor foil and electronics was realized with foil crimps (Crimpflex®

www.dico-electronic.de, accessed on 15 July 2021, DICO Electronic GmbH, Rotenbergstraße 1A, 91126 Schwabach, Germany). On the film side, the crimps were directly connected to the silver tracks of the film, and a housing was plugged over the contact terminals. The mating connector was also connected to a flat ribbon cable by means of crimp contacts and a corresponding housing. The second side of the ribbon cable was connected directly to the readout electronics via terminal plugs. The three foil elements have been laminated onto the ski (Atomic Redster G7; length: 1.82 m; radius: 19.6 m) with a high-performance adhesive tape of the 3M™ VHB™ series (VHB 5952) with a thickness of 1.1 mm (Figure 1b).

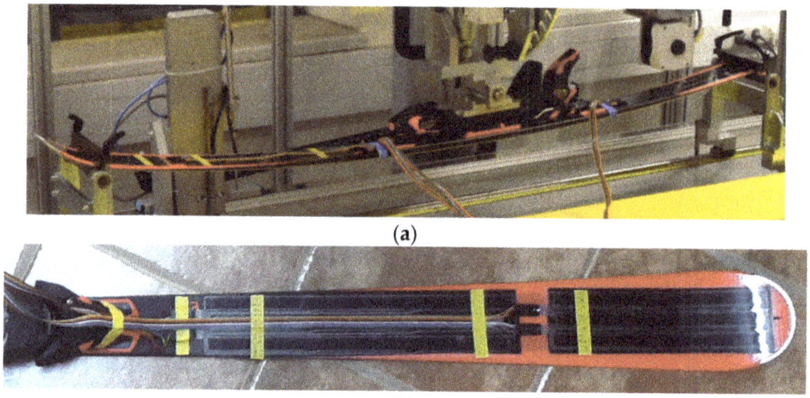

Figure 1. (**a**) Ski prototype with 24 single connected sensors. One foil element was implemented at the rear (nine sensors) and two foil elements at the front (fifteen sensors). (**b**) Detailed view of the sensors in the front ski segment. The sensors were laminated to the ski with a black high-performance adhesive tape (very temperature and UV stable).

A development kit (Joanneum Research, Graz, Austria) was used as a data logger to record the charge response of the PyzoFlex® sensors. It was comprised of a Raspberry Pi 3B+ carrying a self-developed hock-up board containing the analog input stages and a multichannel analog to digital converter (ADC). The Raspberry PI communicates with the ADC (23.7 kSPS/channel) on the hock-up board via a serial peripheral interface. A total of 16 optional sensor signals (out of 24) can be read upon amplification by an operational amplifier circuit. To this end, a charge amplifier circuit (charge-to-voltage converter) with a feedback capacity of $C_f = 20$ nF was chosen. The value of this capacitance and the resulting measuring range were obtained by test measurements on the bending machine (see experimental setup below). The sampling frequency of the system was 215 Hz/channel. The data were transferred via Ethernet to an external device on which self-developed software was used to display the data and store it in an SQLite database.

2.2. Development of the Empirical Curvature Model

To render the printed sensor elements piezoelectric, a poling step is needed [22]. During electrical poling, a high voltage signal is applied to the top and bottom electrodes, which aligns the microscopic dipoles of the ferroelectric layer toward the external electric field. As a result of this poling step, the ferroelectric layer possesses a remnant polarization P_r with an out-of-plane orientation, i.e., normal to the electrodes' surfaces (indexed with (3). The value of P_r is obtained directly from the poling hysteresis and serves as a benchmark for the piezo- and pyroelectric response of the sensor layer. A variation in the average out-of-plane strain component s_{33} directly leads to a change in the polarization, which can be measured as a charge response Q. This electromechanical coupling is described by the piezoelectric constant e_{33} as follows [23,24]:

$$e_{33} = \frac{-Q}{A} \cdot \frac{1}{s_{33}} = -a \cdot P_r \quad (1)$$

where A is the interface area between the electrode and the ferroelectric material, and a is a material constant close to unity.

To derive the bending response of the PyzoFlex® sensors in the ski application, the local bending at the sensor position needs to be related to the induced strain component s_{33}. The scheme in Figure 2a depicts the simplified situation with the sensor clamped on the ski. Bending of the ski with a mean bending radius R causes a lateral, in-plane strain s_{11} in the piezoelectric layer of

$$s_{11} = \zeta / R, \quad (2)$$

with ζ denoting the distance of the ultrathin piezoelectric layer from the neutral axis. The introduced lateral strain translates into a vertical, out-of-plane strain $s_{33} = -\nu_{13} \cdot s_{11}$ with Poisson's ratio ν_{13}. By combining Equations (1) and (2), the bending charge response is derived as:

$$Q = -A \cdot \nu_{13} \cdot a \cdot P_r \cdot \zeta \cdot 1/R. \quad (3)$$

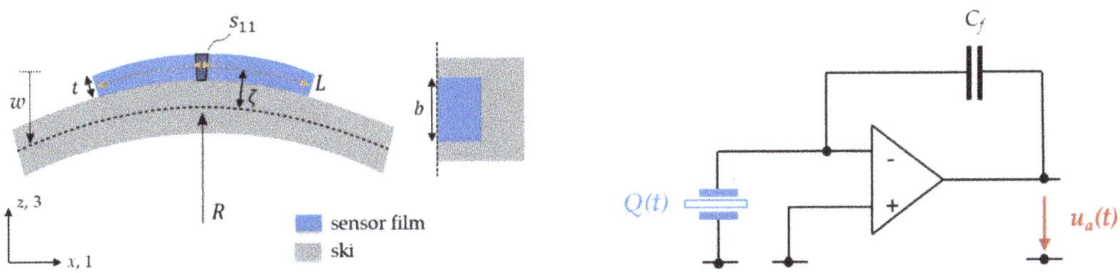

Figure 2. (**a**) The bending scheme of a PyzoFlex® sensor film mounted on top of the ski at the bending radius $R = 1/w''(x)$ (see main text) in cross-section view (left) and top view (right). The bending of the sensor causes a lateral, in-plane strain s_{11} in the piezoelectric layer. The sensor element has length L, width b and thickness t, and the sensitive, piezoelectric layer is located at a radial distance ζ off the neutral axis. (**b**) The sensor generates a charge $Q(t)$, which is converted into a proportional output voltage $u_a(t)$. C_f corresponds to the capacitance in the feedback loop.

Thus, for ideal bending conditions, the charge response is indirectly proportional to the bending radius and directly proportional to the mean curvature $\overline{w_i''(x)} = 1/R$ (Equation (3). It should be noted that ν_{13} and a are intrinsic material properties and therefore equal for all sensor elements, whereas $A = L \cdot b$ depends on the sensor element's geometry (length L and width b, cf. Figure 2a). Finally, P_r can be obtained from the poling process and varies only slightly among the sensor elements. The only unknown factor is the location of the neutral axis, as it depends on the inner structure and elastic properties of the rather complex ski construction. Furthermore, the ski's cross-section and mechanics vary along the ski axis, which results in a position-dependent $\zeta(x)$. Consequently, to be able to derive the local ski bending curvature from the measured charge response, a calibration procedure is necessary for every combination of ski design, sensor design and alignment, respectively.

The input charge $Q(t)$ is converted by a charge amplifier circuit (Figure 2b) into a corresponding output voltage $u_a(t)$. The amplification is determined by C_f in the feedback loop. The output voltage is given as

$$u_a(t) = -\frac{Q(t)}{C_f} = \frac{A \cdot \nu_{13} \cdot a \cdot P_r \cdot \zeta}{C_f} \cdot \overline{w''}(t). \quad (4)$$

No generalized correlation between the PyzoFlex® sensor signal and the curvature could be found in preliminary experiments, as was expected due to the unknown parameter $\zeta(x)$. Instead, a position-specific correlation between $\overline{w''}(x_i)$ and $u_a(x_i, t)$ for the sensor element at position x_i was found. Therefore, each sensor was calibrated under two extreme load conditions. For each sensor a linear regression model was created to derive the mean curvature from the signal in accordance with the physical model (Equation (4)). As calibration factors, the slope k and the intercept d of the linear function of the form

$$\overline{w''}(u_a) = \frac{-C_f}{A \cdot \nu_{13} \cdot a \cdot P_r \cdot \zeta} \cdot (u_a - u_0) = k \cdot u_a + d \quad (5)$$

were used, where u_0 denotes a voltage (i.e., charge) baseline offset.

2.3. Reliability and Validity Assesment

2.3.1. Experimental Setup

In order to deform the ski in a standardized way and measure the deflection with high precision, a three-point bending test was carried out on a bending machine (Atomic GmbH, Austria). The basic structure is a welded steel construction with a solid granite block, which ensures that the deflection measurement can be carried out free from ambient oscillations and vibrations (Figure 3a). The ski was mounted in the front (support A) and rear (support B) end with rotatable clamps about the y-axis. The force F was applied via a trapezoidal spindle linear actuator (Type LMR 03, Servomech S.p.a, Italy) in the middle of the binding system and was monitored by a force sensor (HI U9C, HBM GmbH, Germany). To measure the vertical displacement of the ski, the bending device was equipped with a laser measuring system (LK-H157, Keyence AG, Japan). The laser manufacturer specifies a measuring range of ± 40 mm and a repeatability accuracy of 0.25 µm. The laser unit was moved in x-direction by a servo motor (EMGA-60-P-G5-SAS-70, Festo SE & Co. KG, Germany) with an increment of 20 mm and captured the vertical distance (z-axis) to the ski's underside at each of these points. Since the ski was supported at the front and rear end and the laser unit cannot reach the support ends, the rear 76 mm and the front 190 mm of the 1820 mm ski were not captured. To obtain $w(x)$ over the entire length between the two supports, three data points in the rear and two data points in the front are extrapolated by fitting a second-order polynomial function through the last four measurement points. Consequently, N = 83 data points (78 directly by the laser measurement system and 5 extrapolated) were captured over the ski length of 1650 mm for each deflection (Figure 3b).

To determine the precision of the laser measurement system for the specific experimental setting, the standard deviation (SD) at each measurement point was calculated over 10 repetitions. The maximum SD in the z-direction (laser) was 99.26 µm and in the x-direction (servo motor) 29.88 µm.

Reliability and validity assessments were performed on the instrumented ski described in 2.1, excluding the three sensors in the front since they are located outside support B, and thus, no laser data are available. The remaining 21 sensors were captured in two rounds since the data acquisition device has only 16 inputs; therefore, the measurements were first performed with the outer sensor rows (14 sensors); subsequently, data of the middle sensor row (7 sensors) were recorded. For the study, the ski was deformed with five different load levels and three repetitions at each level (100 N, 110 N, 160 N, 220 N and 230 N). A load of 230 N seems to be sufficiently large, since the vertical deflection in the middle segment is about 60 mm. In comparison, the vertical deformation in exemplary field tests of skis instrumented with strain gauges was a maximum of 40 mm [15–17]. Due to the high precision of the laser and the high time required, laser data was only recorded for the measurements of the outer sensor rows. Considering the symmetrical design of the ski and the pure bending load (no torsion), it is assumed that parallel arranged sensors along the y-axis are deformed to the same amount.

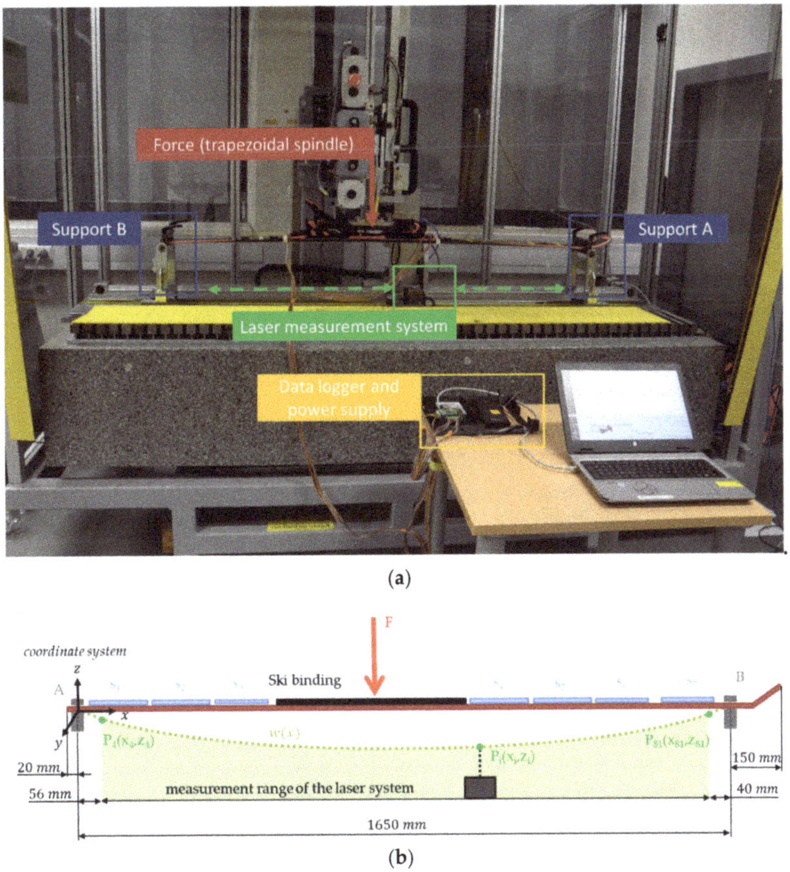

Figure 3. Three-point bending test with an integrated laser measurement system (Atomic GmbH); (**a**) picture with the corresponding experimental components; (**b**) schematic drawing of the experimental setup. N = 83 data points were captured over a length of 1650 mm (78 points directly by the laser measurement system ($P_4(x_4, z_4)$ to $P_{81}(z_{81}, z_{81})$) and 5 points extrapolated).

2.3.2. Data Processing–Laser Measurement System

For every load level, the average value of each measurement point was calculated over the three repetitions. Based on the detected laser data, the curvature function of the ski was derived. In the ski shape model of Yoneyama et al. [17], the vertical displacement was captured with an increment length of 100 mm along the ski length, and a 10th order polynomial function was used for the interpolation. Due to our smaller increment length of 20 mm and the high accuracy of the laser measurement system, a polynomial function of a higher degree seems to be appropriate to represent the curvature progression in a highly differentiated way, but a polynomial function with a higher order causes oscillation in the edge regions due to Runge's phenomenon [25]. This problem can be avoided by using cubic spline curves, which are piecewise polynomials of the third order. The cubic spline minimizes the Holladay functional ($I(w) = \int_a^b (w''(x))^2 \cdot dx$), which causes the second derivative to be best approximated. The spline result can, therefore, be seen as an attempt to interpolate $w(x)$ in such a way that minimal curvatures occur between the data points. A cubic spline interpolation through all 83 data points would fit $w(x)$ very well but generates fluctuations in the second numerical derivative and thus increases the relative error in curvature calculation. Therefore, it is not appropriate to lay splines through all 83 data points. Since the sensors with a length of 100 mm detect an average curvature in the respective segment, it seems reasonable to represent the curvature characteristic of the 1820 mm long ski with the help of 18 splines. Therefore, a cubic spline function with 18 splines was placed through the measuring points to interpolate $w(x)$. By calculating the second spatial numerical derivative of $w(x)$, the curvature function of the ski $w''(x)$ can be obtained:

$$w''(x) \stackrel{\text{def}}{=} \frac{\partial^2 w}{\partial x^2} \tag{6}$$

In order to perform a numerical computation of the mean curvature $\overline{w_i''}$ of the different sensor segments, s_i, $w''(x)$ was interpolated again with cubic splines so that $\overline{w_i''}$ could be calculated over 10 points. The average value of the spline curve over the interval (b_i, a_i) (Figure 4) defines the mean curvature:

$$\overline{w_i''} \stackrel{\text{def}}{=} \overline{\frac{\partial^2 w_i}{\partial x^2}} = \frac{1}{b_i - a_i} \sum_{x=a_i}^{b_i} w''(x) \cdot \Delta x \tag{7}$$

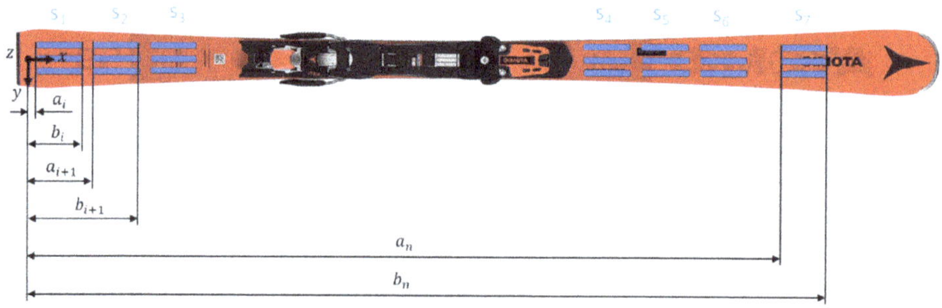

Figure 4. The ski instrumented with PyzoFlex® technology-based sensors.

For each of the seven segments, $\overline{w_i''}$ was calculated. It is assumed that the curvature is constant along the ski width, which means parallel sensors along the y-axis are deformed to the same amount.

2.3.3. Data Processing—PyzoFlex® Sensor System

All PyzoFlex® data recorded during the measurement was calibrated with the empirical curvature model described in Section 2.2. The calibration coefficients k and d (see Equation (5)) were determined using two separate trials at 100 N and 230 N. A 2nd order Butterworth filter was implemented to remove noise from the raw data. In particular, low-frequency and temperature-related signal fluctuations due to the pyroelectric effect are filtered. The high-pass filter with a cut-off frequency of 0.05 Hz was implemented using MATLAB (R2018B, MathWorks, Natick, MA, USA). As the maximum value of the sensor signal is decisive in data-processing, this value was determined by a custom peak detection algorithm in MATLAB.

2.4. Statistical Analysis

To determine the performance of the PyzoFlex® sensors, the aspects of instrument reliability (within the PyzoFlex® sensor system) and criterion validity (between systems) were investigated. For the statistical analysis, all 21 sensors from segments 1 to 7 were included.

For the reliability assessment, the data were checked statistically (Shapiro-Wilk test) for normality of distribution. Changes in the mean (CIM) of $\overline{w''}$ for three load levels (100 N, 160 N and 230 N) were analyzed by the use of a paired sample t-test (level of significance $p < 0.05$). The intra-class correlation coefficient (ICC 3.1) and the typical error of measurement, expressed as the coefficient of variation (CV), were used to assess relative and absolute test-retest reliability. ICC 3.1 was interpreted as not acceptable ($\leq$0.74), good (0.75–0.89) and excellent ($\geq$0.9) [26]. For CIM, CV and ICC 3.1, successive pairs of trials were considered (repetition 1 with repetition 2, repetition 2 with repetition 3).

For the criterion validity between PyzoFlex® and the laser system, Pearson's correlation coefficient was used. All load levels were considered for the statistical evaluation. Magnitudes of correlations were rated as $r < 0.45$, impractical; 0.45–0.70, very poor; 0.70–0.85, poor; 0.85–0.95, good; 0.95–0.995, very good; and >0.995, excellent [27]. Additionally, a Bland–Altman plot was used to describe the agreement between the two quantitative measurement methods [28]. To assess the limits of agreement (LoA) ($\pm$0.96 SD), the accuracy, expressed by the systematic bias, and the precision, expressed by the SD, were calculated.

The paired sample t-test statistics were performed with IBM SPSS Statistics V.26.0 (SPSS Inc., Chicago, IL, USA), and the other metrics were calculated with a custom spreadsheet [27].

3. Results

3.1. Descriptive Report

Figure 5 shows the raw data of the laser measurement system at different load levels (100 N, 110 N, 160 N, 220 N and 230 N). Therefore, the mean $\pm$1.96 SD of each measurement point was calculated over three repetitions. Figure 5a shows the vertical displacement of all 83 measurement points, and Figure 5b a detailed view over 10 measurement points.

The segmental mean curvature (mean $\pm$SD) calculated from the calibrated PyzoFlex® data of the seven middle row sensors (orange and blue points) and the calculated curvature characteristic determined from the data of the laser measuring instrument is represented in Figure 6. The gray large data points correspond to the result of the numerical derivation of the cubic spline function from Equation (6), and the small gray data points are the interpolated points in between. The highest curvature value is observed at sensor position 1 in the rear ski segment. At a load of 230 N, the maximum curvature is 0.26 m^{-1}, which is equivalent to a segmental mean radius of 3.85 m. The vertical displacement in the center of the ski is the highest, as shown in Figure 5, but due to the larger bending stiffness in the binding area, the ski is only slightly curved in this region ($\overline{w''} = 0.11\ m^{-1}$ ($R = 9.09$ m) at 230 N).

3.2. Instrument Reliability and Accuracy

Table 1 shows test-retest reliability results for both repetition pairs at three different load levels (100 N, 160 N and 230 N). The CIM is between -1.41% ($-3.30 \times 10^{-3}\ m^{-1}$) and 0.50% ($0.59 \times 10^{-3}\ m^{-1}$) (over all comparisons). Significant differences are detectable in two of the six repetition comparisons. The maximum CV value of 1.45% can be noted when comparing repetition 1 and 2 at load level 230 N. ICC 3.1 was very close to 1 (ICC 3.1 > 0.961, $p < 0.001$) for all cases.

3.3. Criterion Validity

In Figure 7a, data from the PyzoFlex® sensor system was correlated to the criterion instrument (laser measurement system) to determine validity. The slope of the function as well as the correlation coefficient R^2 are very close to 1. Figure 7b shows the Bland–Altman plot. The systematic bias was $1.33 \times 10^{-3}\ m^{-1}$, and SD was $4.14 \times 10^{-3}\ m^{-1}$. The dotted horizontal lines are at $\pm$1.96 SD and represent the LoA. The points marked in red outside LoA originate from the sensors at segment 7.

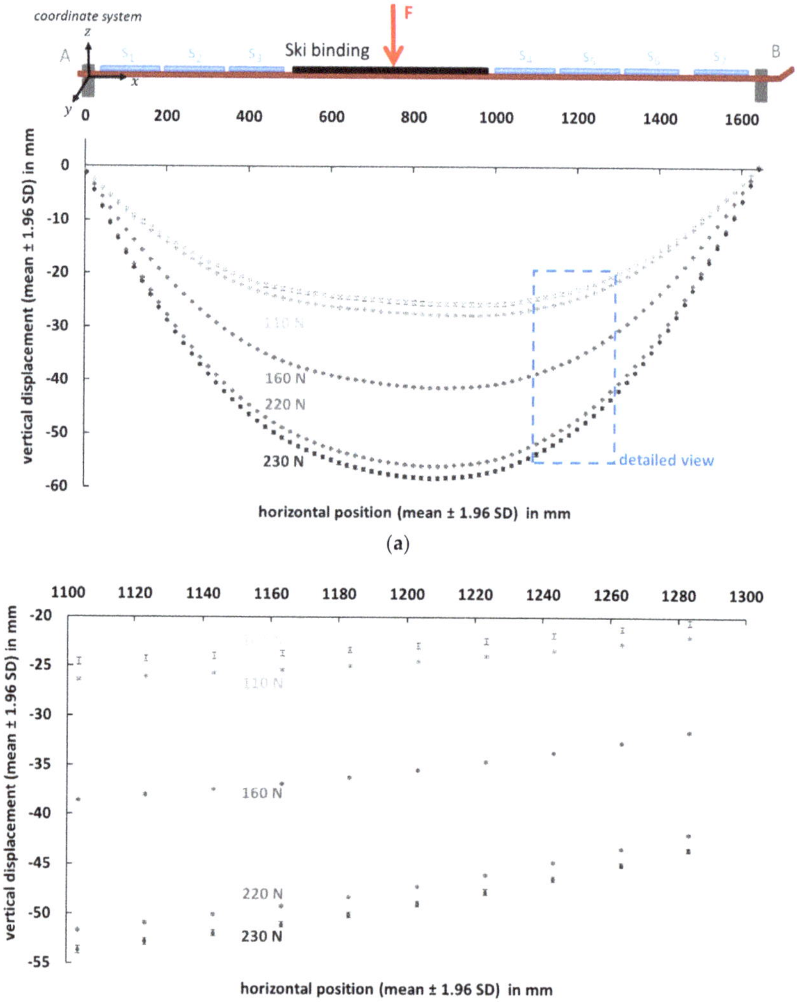

Figure 5. Raw data from the laser measurement system. Vertical displacement (mean ±1.96 SD) vs. horizontal position (mean ±1.96 SD) over three repetitions at different load levels (100 N, 110 N, 160 N, 220 N and 230 N). (**a**) All 83 measuring points; (**b**) Detailed view over 10 measuring points.

Table 1. Analysis of instrument reliability from the sensor system.

		p (*t*-Test)	CIM (%, 95% CI)	CV (%, 95% CI)	ICC 3.1 (95% CI)
100 N	Repetition 1 vs. 2	n.s.	0.50 (−0.08–1.07)	1.08 (0.86–1.47)	0.982 (0.962–0.991)
	Repetition 2 vs. 3	n.s.	0.39 (0.05–0.73)	0.63 (0.51–0.86)	0.994 (0.987–0.997)
160 N	Repetition 1 vs. 2	<0.01	−0.46 (−0.70–0.23)	0.44 (0.35–0.60)	0.995 (0.990–0.998)
	Repetition 2 vs. 3	n.s.	−0.26 (−0.80–0.27)	1.01 (0.81–1.37)	0.977 (0.952–0.989)
230 N	Repetition 1 vs. 2	n.s.	0.32 (−0.45–1.09)	1.45 (1.16–1.97)	0.961 (0.920–0.982)
	Repetition 2 vs. 3	<0.001	−1.41 (−1.86–0.95)	0.86 (0.69–1.17)	0.983 (0.966–0.992)

CIM: change in mean; CI: confidence interval; CV: coefficient of variance; ICC 3.1: intraclass correlation coefficient.

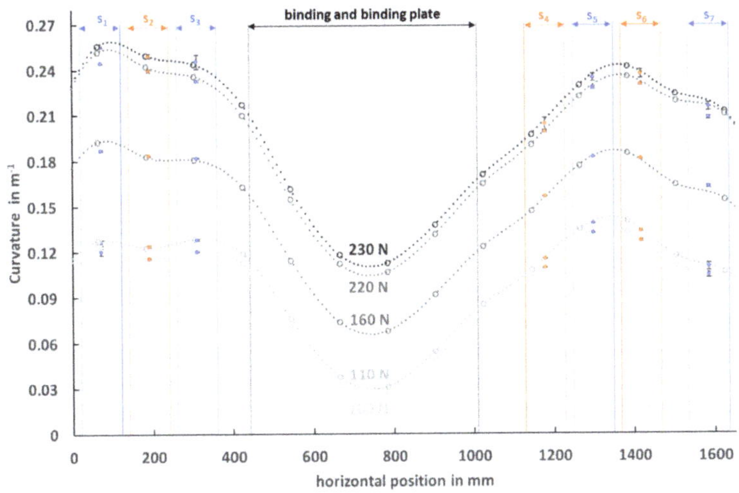

Figure 6. The orange and blue data points represent the calculated mean curvature from the PyzoFlex® data (mean ± SD) of the middle sensor row for the corresponding segments (S_1 to S_7). The large gray data points represent the curvature progression calculated from the laser data at different load levels (100 N, 110 N, 160 N, 220 N and 230 N). The interpolated points (small gray data dots) were used for the numerical calculation of the mean segmental curvature. Note: There are no sensors in the binding area, and for the sake of completeness, the data from the laser measurement system are displayed.

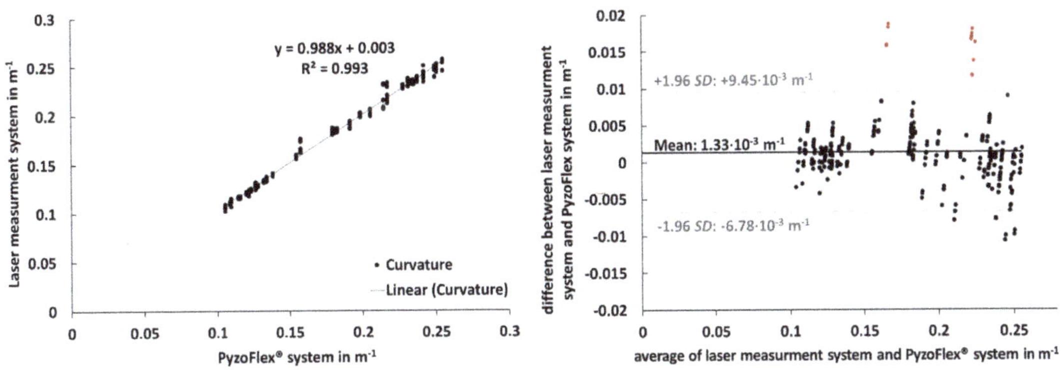

(a) (b)

Figure 7. Left (**a**): The correlation between curvature (m^{-1}) measured by laser measurement system (criterion instrument) and PyzoFlex® sensor system. Right (**b**): Bland–Altman plot showing the difference against the average of the laser measurement system and PyzoFlex® sensor system with limits of agreement (LoA) (dotted lines). SD: Standard deviation.

4. Discussion

The purpose of this study was to develop a novel ski prototype for the detection of segmental ski curvature and to evaluate its test-retest reliability and criterion validity. The PyzoFlex®-Ski-Prototype could be considered as an accurate and precise instrument since data showed a good association between the reference instrument (laser measurement system) and the sensor system under standardized laboratory conditions.

It was shown that the ski demonstrator provides highly reliable data. Both the CIM and CV were below 1.45%. Furthermore, the ICC 3.1 was consistently close to 1 (all ICC 3.1 > 0.961) and indicated excellent reliability [26]. Two of the six repetition comparisons were significantly different, with the highest significance recorded at 230 N when comparing repetition 2 to 3 (CIM = −1.41%, $p < 0.001$). An analogous reliability test of the laser raw data (vertical displacement) shows that all six repetition comparisons differed extremely little in absolute terms but statistically significantly from each other. For the largest difference mentioned (230 N, repetition 2 vs. 3), the CIM (95% CI) for the laser data was −0.05% (−0.03–−0.06) ($p < 0.001$) and the CV (95% CI) is −0.06% (−0.05–−0.07). The minimal but significant differences in the sensor data speak to the accuracy as the differences are attributable to the laser data.

The curvature calculated from the sensor signal is in good agreement with the data of the high-precision laser measurement system. The systematic bias was less than $1.33 \times 10^{-3}\ \mathrm{m}^{-1}$ and therefore shows high accuracy. Additionally, the precision of the signal, which is determined by the limits of agreement ($+9.45 \times 10^{-3}\ \mathrm{m}^{-1}$ and $-6.78 \times 10^{-3}\ \mathrm{m}^{-1}$), indicates a high validity. The results from this study were in line with the coefficient of determination ($R^2 = 0.993$). But it is noticeable that the precision is systematically lower for the outer segment S_7. All points marked in red that are above the LoA line in Figure 7b are related to sensors of segment S_7. Perhaps these small but systematic deviations are due to the extrapolation of data points in the edge region during the laser measurements. This hypothesis is supported by the fact that also at the rear segment S_1, comparatively larger differences are recognizable (Figure 6). However, there may be other reasons as well.

In previous studies, little attention was paid to the ski curvature. Most authors have dealt with the sensor-based detection of the deflection line $w(x)$ but not with $w''(x)$. In those studies, the sensors were calibrated in the laboratory, but there were no reliability and validity tests indicated. Consequently, no comparison to previously used systems can be drawn. Only one study reported that the measurement system was validated, and the RMS error of the ski shape estimation was 0.011 m in the lab [14].

The PyzoFlex®-Ski-Prototype presented here has no sensors in the binding area, which also applies to comparable prototypes [14,18–20]. There is one ski described that is also capable to detect the deflection in the area of the binding system but using a very complex construction [5,15–17]. This design affects the dynamic behavior of the skis [17] and is therefore not suitable for commercial applications (e.g., smart skiing equipment). Of course, in principle, it would be advantageous to acquire sensor data also from the middle segment of the ski, but due to the high bending stiffness, the curvatures occurring in this area are comparatively smaller (see Figure 6). Furthermore, if a skier goes into a backward or forward-leaning position while carving, the curvature will primarily discriminate in the front and rear ski areas and less in the binding area. The skier has no possibility to address only the deflection of the middle ski segment. Therefore, if the deflection of the front and rear ski segments is known, the curvature of the middle segment can be approximated. The integration of the sensor foils into the ski structure over the entire length of the ski might be a future perspective to fully cover the deflection behavior of an entire ski. In the current prototype, it was only possible to attach the sensors to the ski surface. This hypothesis still needs to be checked.

Until now, there are no commercial applications that measure ski deflection during skiing. The ski prototype based on PyzoFlex® technology opens a potential field of application. Not the least because of the low-cost sensor production by screen printing and the flexible sensor design. This sensor layout can easily be modified with regard to size, shape and arrangement of the sensors. Using a hot press process, it is conceivable that the sensor implementation could be embedded in the manufacturing process of the ski. This process protects the sensors from external influences, and sensors can also be integrated under the binding area. Another advantage over strain gauge-based systems is that the film not only functions as a sensor, but also as a mechanical-electrical energy converter. The film not only functions as a sensor, but also as a mechanical-electrical energy converter. Combined with suitable harvester electronics, an area-related power in the range of several $\mu W \cdot cm^{-2}$ can be generated [22]. There is already a prototype with sensors mounted on the inside of a car tire. At 30–50 $km \cdot h^{-1}$, enough energy is generated to send the measured tire pressure protocol to the board computer every 30 s [22].

It cannot be answered on the basis of these investigations if the used PyzoFlex sensors are generally more suitable for the curvature measurement compared to other sensor technologies because each sensor system has its advantages and disadvantages. However, it should be mentioned that not one of the various prototypes or sensor systems has succeeded in reaching the market, and there are reasons for this. However, the main difference from comparable studies is that the method of deformation analysis is a completely new approach. The subjective judgement leads to the belief that a bent ski is most curved in the middle. This is a fallacy, as the vertical deflection has little to do with the localization of the highest curvature. The curvature progression of the ski and its change over time provides insights into the quality of a turn. If only the bending line is analyzed, this information cannot be extracted, and this distinguishes this work from the others. To determine segmental curvature maxima (or local radius minima), the second derivative of the bending line must be investigated.

However, the current study has some limitations. All measurements were performed in a quasi-static setting. Therefore, a transfer into the field (i.e., on-snow skiing) is only possible to a certain extent. In skiing, there are multifactorial parameters that influence ski deflection. For example, the influence of ski vibrations different excitation frequencies and torsional deformations on the sensor signal cannot be stated yet based on this study. The sensor layout was chosen in such a way that differences of parallel sensors can be used to detect the segmental torsion angles along the ski. This requires an extension of the bending machine to be able to perform combined loads (torsion and bending). Furthermore, the influence of changes in temperature is not yet clearly defined. In principle, dynamic temperature changes can substantially influence the sensor signal due to the pyroelectric effect. So far, the temperature was assumed to be quasi-static because it changes sufficiently slowly (e.g., the heat generated by ski deformations); therefore, temperature changes are predominantly in the low-frequency range, which means that they can be filtered out well with a high-pass filter. Absolute temperature differences do influence the piezoelectric coupling coefficient, but only very weakly [29]. This dependence must be analyzed in more detail in order to introduce a temperature-dependent calibration factor if necessary. Further measurements and analyses are required to answer these open questions.

5. Conclusions

The deflection of the ski is the result of the ski-snow interaction and provides information about the quality of a turn. Until now, there are no commercial applications that measure ski deflection during skiing.

A new ski prototype has been developed to determine the segmental curvature of Alpine skis. The PyzoFlex® technology-based sensor foils are a valid and reliable measuring instrument to measure the segmental curvature characteristic of alpine skis in a standardized setting.

Validating the curvature model in the field is complex since we do not know of any gold standard measuring instrument on snow. A dynamic validation in the laboratory would be a possibility (e.g., by a programmable bending robot) to apply deformations similar to skiing (cyclic deformation, edging). In perspective, not only cyclic loads but also combined loads (torsion and bending) could be applied to determine the scope of the curvature model and to extend it if necessary. Another approach would be to perform a technical validation in the field. However, further improvements of the prototype are planned for the future (e.g., waterproofing, wireless transmission of sensor data, integration of the sensor foils into the ski structure to protect them).

Author Contributions: Conceptualization, C.T., J.K. and T.S.; methodology, C.T., J.K., A.T. and P.S.; validation, J.K. and T.S.; formal analysis, J.K. and T.S.; investigation, C.T. and A.T.; resources, A.T., and H.H.; data curation C.T.; writing—original draft preparation, C.T., A.T. and P.S.; writing—review and editing, P.S., J.K. and T.S.; visualization, C.T.; supervision, J.K. and T.S.; project administration, T.S.; funding acquisition, J.K. and T.S. All authors have read and agreed to the published version of the manuscript.

Funding: This work was partly funded by the Austrian Federal Ministry for Transport, Innovation and Technology, the Austrian Federal Ministry for Digital and Economic Affairs, and the federal state of Salzburg, as well as from the European Union's Horizon 2020 research and innovation program.

Institutional Review Board Statement: Not applicable.

Informed Consent Statement: Not applicable.

Data Availability Statement: The data presented in this study are available on request from the corresponding author.

Acknowledgments: The authors would like to thank Atomic Austria GmbH for the good cooperation.

Conflicts of Interest: The authors declare no conflict of interest.

References

1. Lieu, D.K.; Mote, C.D. Mechanics of the turning snow ski. In *Skiing Trauma and Safety: Fifth International Symposium*; Johnson, R.J., Mote, C.D., Eds.; ASTM International: Philadelphia, PA, USA, 1985; Volume 1, pp. 117–140.
2. Renshaw, A.A.; Mote, C., Jr. A model for the turning snow ski. *Int. J. Mech. Sci.* **1989**, *31*, 721–736. [CrossRef]
3. Reid, R.C.; Haugen, P.; Gilgien, M.; Kipp, R.W.; Smith, G.A. Alpine Ski Motion Characteristics in Slalom. *Front. Sports Act. Living* **2020**, *2*, 25. [CrossRef] [PubMed]
4. Brown, C.A.; Outwater, J.O. On the skiability of snow. In *Skiing Trauma and Safety: Seventh International Symposium*; Johnson, R.J., Mote, C.D., Binet, M.H., Eds.; ASTM International: Philadelphia, PA, USA, 1989; Volume 1, pp. 329–336.
5. Kagawa, H.; Yoneyama, T.; Tatsuno, D.; Scott, N.; Osada, K. Development of a measuring system on ski deflection and contacting snow pressure in turns. In *Science and Skiing IV*; Müller, E., Lindinger, S., Stöggl, T., Eds.; Meyer & Meyer Sport: Aachen, Germany, 2009; Volume 1, pp. 281–291.
6. Nachbauer, W.; Rainer, F.; Schindelwig, K. Effects of ski stiffness on ski performance. In Proceedings of the Engineering of Sport 5, Winfield, IL, USA, 13–16 September 2004; pp. 472–478.
7. Rainer, F.; Nachbauer, W.; Schindelwig, K.; Kaps, P. On the measurement of the stiffness of skis. In *Science and Skiing III*; Müller, E., Bacharach, D., Klika, R., Eds.; Meyer & Meyer Sport: Aachen, Germany, 2005; Volume 1, pp. 136–147.
8. Deak, A.; Jorgensen, J.; Vagners, J. The engineering characteristics of snow skis—part 1: Static bending and torsional characteristics. *J. Eng. Ind.* **1975**, *97*, 131–137. [CrossRef]
9. Lüthi, A.; Federolf, P.; Fauve, M.; Rhyner, H. Effect of bindings and plates on ski mechanical properties and carving performance. In *The Engineering of Sport 6*; Hubbard, M., Metha, R.D., Pallis, J.M., Eds.; Springer: Winfield, IL, USA, 2006; Volume 1, pp. 299–304.
10. De Gobbi, M.; Petrone, N. Structural behaviour of slalom skis in bending and torsion (P269). *Eng. Sport 7* **2008**, *2*, 643–652.
11. Truong, J.; Brousseau, C.; Desbiens, A.L. A Method for Measuring the Bending and Torsional Stiffness Distributions of Alpine Skis. *Procedia Eng.* **2016**, *147*, 394–400. [CrossRef]
12. Clifton, P.; Subic, A.; Mouritz, A. Snowboard stiffness prediction model for any composite sandwich construction. *Procedia Eng.* **2010**, *2*, 3163–3169. [CrossRef]
13. International Organization for Standardization. *Alpine skis—Determination of the Elastic Properties*. (ISO Standard No. 5902:2013). 2013. Available online: https://www.iso.org/standard/60844.html (accessed on 15 July 2016).
14. Adelsberger, R.; Aufdenblatten, S.; Gilgien, M.; Tröster, G. On bending characteristics of skis in use. *Procedia Eng.* **2014**, *72*, 362–367. [CrossRef]
15. Yoneyama, T.; Kitade, M.; Osada, K. Investigation on the ski-snow interaction in a carved turn based on the actual measurement. *Procedia Eng.* **2010**, *2*, 2901–2906. [CrossRef]
16. Tatsuno, D.; Yoneyama, T.; Kagawa, H.; Scott, N.; Osada, K. Measurement of ski deflection and ski-snow contacting pressure in an actual ski turn on the snow surface. In *Science and Skiing IV*; Müller, E., Lindinger, S., Stöggl, T., Eds.; Meyer & Meyer Sport: Aachen, Germany, 2009; Volume 1, pp. 505–515.
17. Yoneyama, T.; Scott, N.; Kagawa, H.; Osada, K. Ski deflection measurement during skiing and estimation of ski direction and edge angle. *Sports Eng.* **2008**, *11*, 3–13. [CrossRef]
18. Schindelwig, K.; Hellberg, F.; Nachbauer, W.; Schretter, H. Measuring dynamic ski behavior with strain gauges. In Proceedings of the ISBS—Conference (XXIV Symposium on Biomechanics in Sports), Salzburg, Austria, 14–18 July 2006; p. 1.
19. Fauve, M.; Auer, M.; Lüthi, A.; Rhyner, H.; Meier, J. Measurement of dynamical ski behavior during alpine skiing. In *Science and Skiing IV*; Müller, E., Lindinger, S., Stöggl, T., Eds.; Mayer & Mayer Sport: Aachen, Germany, 2009; Volume 1, pp. 195–206.

20. Kos, A.; Umek, A. Smart sport equipment: SmartSki prototype for biofeedback applications in skiing. *Pers. Ubiquitous Comput.* **2018**, *22*, 535–544. [CrossRef]
21. Rendl, C.; Greindl, P.; Haller, M.; Zirkl, M.; Stadlober, B.; Hartmann, P. PyzoFlex: Printed piezoelectric pressure sensing foil. In Proceedings of the 25th Annual ACM Symposium on User Interface Software and Technology, Cambridge, MA, USA, 7–10 October 2012; pp. 509–518.
22. Stadlober, B.; Zirkl, M.; Irimia-Vladu, M. Route towards sustainable smart sensors: Ferroelectric polyvinylidene fluoride-based materials and their integration in flexible electronics. *Chem. Soc. Rev.* **2019**, *48*, 1787–1825. [CrossRef] [PubMed]
23. Furukawa, T.; Wen, J.; Suzuki, K.; Takashina, Y.; Date, M. Piezoelectricity and pyroelectricity in vinylidene fluoride/trifluoroethylene copolymers. *J. Appl. Phys.* **1984**, *56*, 829–834. [CrossRef]
24. Schäffner, P.; Zirkl, M.; Schider, G.; Groten, J.; Belegratis, M.R.; Knoll, P.; Stadlober, B. Microstructured single-layer electrodes embedded in P (VDF-TrFE) for flexible and self-powered direction-sensitive strain sensors. *Smart Mater. Struct.* **2020**, *29*, 085040. [CrossRef]
25. Runge, C. Über empirische Funktionen und die Interpolation zwischen äquidistanten Ordinaten. *Z. Math. Phys.* **1901**, *46*, 20.
26. Koo, T.K.; Li, M.Y. A guideline of selecting and reporting intraclass correlation coefficients for reliability research. *J. Chiropr. Med.* **2016**, *15*, 155–163. [CrossRef] [PubMed]
27. Hopkins, W.G. Spreadsheets for Analysis of Validity and Reliability. *Sportscience* **2017**, *21*, 36–44.
28. Bland, J.M.; Altman, D.G. Measuring agreement in method comparison studies. *Stat. Methods Med. Res.* **1999**, *8*, 135–160. [CrossRef] [PubMed]
29. Omote, K.; Ohigashi, H.; Koga, K. Temperature dependence of elastic, dielectric, and piezoelectric properties of "single crystalline" films of vinylidene fluoride trifluoroethylene copolymer. *J. Appl. Phys.* **1997**, *81*, 2760–2769. [CrossRef]

MDPI
St. Alban-Anlage 66
4052 Basel
Switzerland
Tel. +41 61 683 77 34
Fax +41 61 302 89 18
www.mdpi.com

Sensors Editorial Office
E-mail: sensors@mdpi.com
www.mdpi.com/journal/sensors

www.ingramcontent.com/pod-product-compliance
Lightning Source LLC
Chambersburg PA
CBHW062032220926
51850CB00010B/2517
9783036570884